How to start a farm business

Welcome to the comprehensive guide on starting a farm business! Whether you're a seasoned agricultural enthusiast looking to expand your operations or someone dreaming of turning a passion for farming into a thriving enterprise, this book is designed to be your trusted companion. In these pages, you'll find a wealth of practical advice, strategic insights, and invaluable tips gathered from years of experience and expertise in the field. From laying the groundwork for your farm's vision to navigating the complexities of market trends, sustainable practices, and financial management, each chapter is crafted to equip you with the knowledge and confidence needed to embark on and sustain a successful farming venture. Get ready to cultivate your dreams and grow a farm business that not only flourishes but also contributes positively to our agricultural community and environment.

Copyright © 2024

All rights reserved. No part of this book may be reproduced in any form or by any electronic or mechanical means, including information storage and retrieval systems, without permission in writing from the publisher, except by a reviewer, who may quote brief passages in a review.

The information contained in this book is for general information purposes only. The information is provided by naciro and while we endeavor to keep the information up to date and correct, we make no representations or warranties of any kind, express or implied, about the completeness, accuracy, reliability, suitability or availability with respect to the book or the information, products, services, or related graphics contained in the book for any purpose. Any reliance you place on such information is therefore strictly at your own risk.

All trademarks and registered trademarks are the property of their respective owners and are used in this book only for identification and explanation.

Permission to use copyrighted material in this book should be obtained from the copyright owner or the publisher.

This book is not intended to provide medical, legal, or financial advice, and the author and publisher specifically disclaim any liability for any loss or damage caused or alleged to be caused directly or indirectly by the information in this book.

Naciro and the publisher of this book do not endorse or recommend any commercial products, processes, or services. The views and opinions of authors expressed in this book do not necessarily state or reflect those of the publisher of this book.

Contents

Chapter 1: Introduction to Farming

Chapter 2: Benefits and Challenges of Starting a Farm Business

Chapter 3: Understanding Different Types of Farming

Chapter 4: Researching and Choosing Your Farming Niche

Chapter 5: Developing a Farm Business Plan

Chapter 6: Securing Land for Your Farm

Chapter 7: Planning and Designing Your Farm Layout

Chapter 8: Choosing Crops and Livestock for Your Farm

Chapter 9: Farm Infrastructure and Equipment

Chapter 10: Farming Methods and Techniques

Chapter 11: Farm Marketing and Sales Strategies

Chapter 12: Farm Financial Management and Planning

Chapter 13: Farming Regulations and Compliance

Chapter 14: Farm Risk Management

Chapter 15: Sustainable Agriculture Practices

Chapter 16: Farm Technology and Innovation

Chapter 17: Marketing and Branding Your Farm Products

Chapter 18: Financial Management for Your Farm

Chapter 19: Sustainable Farming Practices

Chapter 20: Managing Farm Labor

Chapter 21: Marketing Your Farm Products

Chapter 22: Managing Farm Risks

Chapter 23: The Future of Agriculture: Innovations and Trends

Chapter 24: Navigating Farm Regulations and Compliance

Chapter 25: Building a Farm Brand and Story

Chapter 26: Sustainable Farming Practices

Chapter 28: Embracing Technology in Modern Farming

Chapter 29: Farm Diversification - Strengthening Your Agricultural Enterprise

Chapter 30: Financial Management for Farmers

Chapter 1: Introduction to Farming

Welcome to the exciting world of farming! Whether you're drawn to the allure of working the land, passionate about sustainable agriculture, or simply eager to embark on a new entrepreneurial journey, starting a farm business is a fulfilling and impactful endeavor. In this chapter, we'll explore what farming entails, the benefits it offers, and the foundational considerations you need to kickstart your farm business.

What is Farming?

At its core, farming is the practice of cultivating land, raising crops, and/or breeding animals for food, fiber, fuel, or other products essential to human life. It's a timeless profession deeply rooted in human history, evolving from traditional practices to modern, innovative methods that sustainably meet the needs of a growing global population.

The Rewards of Farming

Farming offers a unique blend of personal satisfaction, environmental stewardship, and economic potential. Here are some of the key rewards:

Connection with Nature:

Farming allows you to work closely with the land, observing the seasons, and becoming attuned to natural rhythms. There's a profound satisfaction in nurturing plants or animals and seeing the fruits of your labor flourish.

Sustainability and Environmental Impact:

As a farmer, you have the opportunity to practice sustainable agriculture, which prioritizes soil health, biodiversity, and efficient resource management. By adopting sustainable practices, you contribute positively to environmental conservation and resilience.

Community and Local Economy:

Farming is often deeply integrated into local communities, supplying fresh produce and goods, supporting local markets, and fostering connections between farmers and consumers. It's a chance to contribute to a vibrant local economy and build relationships with neighbors and customers.

Entrepreneurial Freedom:

Starting a farm business provides a level of independence and entrepreneurial freedom. You have the autonomy to make decisions about what to grow or raise, how to market your products, and how to manage your operations. This independence can be incredibly rewarding and allows for creative innovation in farming practices.

Key Considerations Before Starting Your Farm Business

Before you dig into the soil or purchase your first livestock, there are several crucial considerations to keep in mind:

Passion and Commitment:

Farming requires dedication and perseverance. It's important to assess your passion for agriculture and your commitment to the lifestyle and demands of running a farm business.

Research and Education:

Take the time to research different farming methods, explore various agricultural niches, and educate yourself about the specific requirements of the crops or livestock you intend to raise. Joining workshops, attending seminars, or apprenticing on established farms can provide invaluable hands-on experience.

Financial Planning:

Starting a farm business involves significant upfront costs for land, equipment, seeds, livestock, and infrastructure. Develop a detailed financial plan that includes budgeting for initial investments, operational expenses, and potential income streams.

Legal and Regulatory Requirements:

Understanding the legal and regulatory framework for farming in your region is essential. This includes zoning laws, permits, licenses, and compliance with environmental and health regulations related to farming practices and product sales.

Market Research:

Evaluate market demand for your products. Consider who your target customers are, how you will reach them, and what marketing strategies will be effective. Building a strong market presence and customer base is critical for the success of your farm business.

Conclusion

Embarking on a farm business journey is both a challenging and rewarding endeavor. By understanding the fundamentals of farming, recognizing its rewards, and carefully planning your approach, you set a solid foundation for building a successful and sustainable farm business. In the chapters ahead, we will delve deeper into each aspect of starting and managing a farm, offering practical insights and guidance to help you navigate this exciting venture with confidence. So, roll up your sleeves and get ready to cultivate your dreams on the fertile ground of agriculture!

Chapter 2: Benefits and Challenges of Starting a Farm Business

Congratulations on taking the leap into the world of farming! In this chapter, we'll explore the myriad benefits that come with starting a farm business, as well as the challenges you may encounter along the way. Understanding both sides of the coin will help you navigate this exciting journey with confidence and preparedness.

Benefits of Starting a Farm Business

Starting a farm business offers a wealth of benefits that go beyond financial rewards. Here are some of the key advantages:

1. Quality of Life:

Farming provides a lifestyle connected to nature and the outdoors. You'll enjoy fresh air, beautiful landscapes, and the satisfaction of working with your hands to nurture plants or animals.

2. Food Security and Self-Sufficiency:

As a farmer, you have the opportunity to produce your own food or contribute to local food systems. This self-sufficiency can provide peace of mind and a sense of security, knowing where your food comes from and how it's grown.

3. Environmental Stewardship:

Farmers play a crucial role in environmental conservation. Through sustainable farming practices such as crop rotation, soil conservation, and organic methods, you can contribute to preserving natural resources and biodiversity.

4. Community Engagement:

Farming often fosters strong connections within local communities. You'll have the chance to participate in farmers' markets, community-supported agriculture (CSA) programs, and other initiatives that support local economies and promote healthy eating.

5. Entrepreneurial Opportunities:

Starting a farm business allows you to be your own boss and make independent decisions about your operation. You can innovate in farming techniques, diversify your products, and explore niche markets that align with your interests and values.

6. Health and Wellbeing:

Farming can be physically demanding but also incredibly rewarding for your overall health. Engaging in physical work outdoors can reduce stress, promote physical fitness, and improve mental wellbeing.

Challenges of Starting a Farm Business

While the benefits of farming are significant, it's essential to be aware of the challenges you may face:

1. Financial Investment:

Starting a farm requires substantial initial investment in land, equipment, seeds, livestock, and infrastructure. Securing financing and managing cash flow are critical challenges for new farmers.

2. Market Uncertainty:

Farming is influenced by market fluctuations, weather conditions, and consumer demand. Understanding market trends, diversifying your products, and building resilient business strategies are essential for long-term success.

3. Labor Intensity:

Farming can be physically demanding and labor-intensive, especially during peak seasons like planting and harvest. Managing labor resources efficiently and maintaining workforce motivation are ongoing challenges.

4. Regulatory Compliance:

Farmers must comply with a variety of regulations related to land use, environmental protection, food safety, and labor laws. Staying informed about regulatory requirements and maintaining compliance can be complex and time-consuming.

5. Climate and Weather Risks:

Agriculture is inherently vulnerable to climate variability, extreme weather events, and natural disasters. Implementing climate-resilient farming practices and having contingency plans are crucial for mitigating risks.

6. Skills and Knowledge:

Farming requires a diverse set of skills, from crop management to animal husbandry, marketing, and financial management. Continuously learning and adapting to new technologies and practices are essential for farm success.

Conclusion

Starting a farm business is a rewarding journey filled with opportunities to contribute to food security, environmental sustainability, and community well-being. By understanding the benefits and challenges associated with farming, you can approach your venture with realistic expectations and a proactive mindset. In the chapters ahead, we will delve deeper into each aspect of overcoming challenges and maximizing the benefits of starting and managing a farm business. Get ready to embrace the joys and tackle the hurdles of farming as you embark on this fulfilling adventure!

Chapter 3: Understanding Different Types of Farming

Welcome to the diverse world of farming! In this chapter, we'll explore the various types of farming practices and systems you can consider for your farm business. Each type offers unique opportunities and challenges, allowing you to align your interests, resources, and goals with the most suitable approach to farming.

Traditional vs. Modern Farming Methods

Farming methods have evolved significantly over time, influenced by technological advancements, environmental awareness, and market demands. Understanding the distinctions between traditional and modern farming can guide your approach to building a sustainable and productive farm business.

1. Traditional Farming:

Traditional farming methods are rooted in age-old practices passed down through generations. These methods often emphasize manual labor, natural fertilizers, and seasonal farming cycles. Traditional farming may include:

- **Subsistence Farming:** Focuses on producing enough food to meet the needs of the farmer and their family, often practiced in rural or developing regions.
- **Small-Scale Farming:** Involves cultivating a small plot of land with diversified crops and livestock, aiming for self-sufficiency and local markets.
- **Organic Farming:** Avoids synthetic chemicals and emphasizes natural soil fertility and pest management techniques.

2. Modern Farming:

Modern farming incorporates technological innovations and scientific knowledge to maximize efficiency, productivity, and sustainability. Key features of modern farming include:

- **Industrial Agriculture:** Large-scale operations that use machinery, synthetic fertilizers, and pesticides to achieve high yields and meet global demand.
- **Precision Agriculture:** Utilizes GPS, sensors, and data analytics to optimize inputs such as water, fertilizer, and pesticides, reducing waste and environmental impact.
- **Vertical Farming:** Cultivates crops indoors or in controlled environments using stacked layers or hydroponic systems, maximizing space and resource efficiency.

Types of Farming Systems

Beyond methods, farming can be categorized into various systems based on the primary focus of production, environmental conditions, and market orientation. Each farming system offers distinct advantages and considerations for aspiring farmers.

1. Crop Farming:

Crop farming focuses on cultivating plants for food, fiber, fuel, or pharmaceuticals. Common crops include grains, vegetables, fruits, and specialty crops like herbs or flowers. Considerations include soil health, crop rotation, and pest management strategies.

2. Livestock Farming:

Livestock farming involves raising animals for meat, dairy, fiber, or other products. Types of livestock farming include:

- **Dairy Farming:** Produces milk and dairy products from cows, goats, or sheep.
- **Poultry Farming:** Raises chickens, turkeys, ducks, or other birds for meat and eggs.
- **Livestock Ranching:** Grazes cattle, sheep, or goats on extensive pasturelands for meat production.

3. Mixed Farming:

Mixed farming integrates crop and livestock production within the same operation. This approach can provide diversification benefits, such as nutrient cycling, pest control, and income stability.

4. Specialty Farming:

Specialty farming focuses on niche markets or unique products that cater to specific consumer preferences or trends. Examples include organic farming, heirloom vegetables, medicinal herbs, or agri-tourism activities like pumpkin patches or Christmas tree farms.

5. Agroforestry:

Agroforestry integrates trees and shrubs with crops or livestock, promoting biodiversity, soil conservation, and multiple income streams. Agroforestry systems may include alley cropping, windbreaks, or silvopasture.

Choosing the Right Farming Approach for You

Selecting the most suitable farming approach involves considering factors such as your interests, available resources (land, capital, equipment), market opportunities, and environmental conditions. It's essential to:

- **Evaluate Market Demand:** Research local and global market trends to identify profitable opportunities for your chosen farming system.
- **Assess Resource Availability:** Determine the resources needed (land, water, labor) and assess whether you have access to these resources or can acquire them feasibly.
- **Consider Environmental Impact:** Choose farming practices that minimize environmental footprint and promote long-term sustainability.
- **Seek Knowledge and Expertise:** Continuously educate yourself about farming techniques, market dynamics, and technological innovations relevant to your chosen farming approach.

Conclusion

Understanding the different types of farming methods and systems empowers you to make informed decisions about the direction of your farm business. Whether you opt for traditional, modern, or a blend of farming practices, each approach offers opportunities for innovation, sustainability, and economic viability. In the chapters ahead, we will delve deeper into specific farming techniques, management practices, and strategic considerations to help you build and grow a successful farm business. Get ready to cultivate your passion and reap the rewards of farming in the diverse and dynamic agricultural landscape!

Chapter 4: Researching and Choosing Your Farming Niche

Welcome to the exciting phase of researching and choosing your farming niche! In this chapter, we'll explore the importance of selecting the right farming niche, how to conduct effective research, and considerations to help you make an informed decision that aligns with your interests, goals, and market opportunities.

Why Choose a Farming Niche?

Choosing a farming niche is more than just selecting what to grow or raise—it's about identifying a specialized area of focus that sets your farm apart in the market. By specializing in a niche, you can:

- **Meet Market Demand:** Focus on products or services that have high consumer demand and profitability.
- **Differentiate Your Business:** Stand out from competitors by offering unique or specialty products that appeal to specific customer preferences.
- **Optimize Resource Use:** Efficiently utilize your resources (land, labor, capital) to maximize productivity and profitability.
- **Build a Brand:** Establish a reputation for quality and expertise in your chosen niche, attracting loyal customers and enhancing market visibility.

Conducting Market Research

Effective market research is crucial for identifying viable farming niches and understanding consumer preferences and trends. Here are steps to guide your research process:

1. Identify Potential Niches:

Explore different farming niches based on your interests, skills, and local market opportunities. Consider factors such as:

- **Product Demand:** Analyze consumer preferences and market trends for specific crops, livestock, or products.
- **Location:** Assess local climate conditions, soil suitability, and market proximity for different farming options.
- **Personal Interests:** Choose a niche that aligns with your passion and expertise, ensuring long-term commitment and satisfaction.

2. Evaluate Market Demand:

Research market demand for your chosen niche by:

- **Surveying Potential Customers:** Conduct surveys or interviews to gather feedback on product preferences and willingness to purchase.
- **Analyzing Competitors:** Study existing farms and businesses in your niche to understand their market share, pricing strategies, and customer base.

3. Assess Profitability:

Evaluate the financial viability of each farming niche by:

- **Cost Analysis:** Estimate startup costs, ongoing expenses, and potential revenue streams for each niche.
- **Profit Margin:** Calculate potential profitability based on market prices, production yields, and operational efficiency.

4. Consider Sustainability and Scalability:

Assess the sustainability and scalability of your chosen niche by:

- **Environmental Impact:** Evaluate farming practices that promote soil health, water conservation, and biodiversity.
- **Growth Potential:** Determine opportunities for expanding production or diversifying product offerings over time.

Choosing Your Farming Niche

Once you've conducted thorough research, it's time to make an informed decision on your farming niche. Consider the following factors:

1. Passion and Expertise:

Choose a niche that aligns with your interests, skills, and values. Your passion and expertise will drive your commitment and enthusiasm for your farm business.

2. Market Potential:

Select a niche with strong market demand and growth opportunities. Look for niches that have sustainable consumer interest and potential for profitability.

3. Resource Availability:

Assess your resources (land, equipment, capital) and choose a niche that matches your available resources or that you can acquire feasibly.

4. Risk Management:

Consider potential risks and challenges associated with your chosen niche, such as weather variability, market fluctuations, or regulatory changes. Develop contingency plans to mitigate risks.

5. Long-Term Vision:

Plan for the future growth and development of your farm business within your chosen niche. Evaluate how you can innovate, expand, or adapt to changing market conditions over time.

Conclusion

Choosing your farming niche is a pivotal decision that shapes the direction and success of your farm business. By conducting thorough research, evaluating market opportunities, and considering your personal interests and resources, you can confidently select a niche that aligns with your goals and sets your farm up for long-term profitability and sustainability. In the chapters ahead, we will explore strategies for planning and implementing your chosen farming niche, ensuring you have the knowledge and tools to thrive in the competitive agricultural market. Get ready to carve out your unique space in the world of farming and cultivate success with purpose and passion!

Chapter 5: Developing a Farm Business Plan

Welcome to the crucial step of developing a farm business plan! In this chapter, we'll delve into the importance of having a well-structured business plan for your farm, what it should include, and how to create one that sets a clear path for your success in farming.

Why You Need a Farm Business Plan

A farm business plan serves as a roadmap that outlines your farm's goals, strategies, resources, and financial projections. It provides a comprehensive framework to guide your decisions and operations, whether you're starting a new farm or expanding an existing one. Here are key reasons why a farm business plan is essential:

1. Clarity and Direction:

A well-written business plan clarifies your farm's mission, vision, and objectives. It helps you articulate your goals and define the steps needed to achieve them, providing a clear direction for your farm business.

2. Strategic Planning:

Developing a business plan forces you to analyze market opportunities, assess potential risks, and devise strategies for managing resources effectively. It encourages proactive decision-making and strategic thinking.

3. Resource Management:

By outlining your farm's resource needs (land, equipment, labor, capital), a business plan helps you allocate resources efficiently. It ensures that you have the necessary resources to operate and grow your farm sustainably.

4. Financial Planning and Management:

A business plan includes financial projections, budgets, and cash flow forecasts. It helps you estimate startup costs, identify revenue streams, and plan for ongoing expenses, ensuring financial stability and growth.

5. Communication and Accountability:

A business plan serves as a communication tool for stakeholders, including lenders, investors, and partners. It demonstrates your commitment to your farm business and holds you accountable to your goals and timelines.

Components of a Farm Business Plan

A comprehensive farm business plan typically includes the following components:

1. Executive Summary:

- **Overview:** Briefly describe your farm business, its mission, and the goals outlined in the plan.
- **Highlights:** Summarize key points, including market opportunities, competitive advantages, and financial projections.

2. Business Description:

- **Farm Overview:** Provide an in-depth description of your farm, its location, size, and facilities.
- **Legal Structure:** Specify the legal structure (sole proprietorship, partnership, LLC) and ownership details.

3. Market Analysis:

- **Target Market:** Define your target customers and market segments.
- **Industry Analysis:** Assess market trends, customer needs, and competitive landscape in your chosen farming niche.

4. Products and Services:

- **Product Offering:** Detail the products or services your farm will produce (e.g., crops, livestock, value-added products).
- **Unique Selling Proposition (USP):** Highlight what sets your products apart from competitors.

5. Marketing and Sales Strategy:

- **Marketing Plan:** Outline strategies for promoting your farm products, including branding, advertising, and customer outreach.
- **Sales Strategy:** Describe how you will distribute and sell your products, such as direct sales, farmers' markets, or wholesale partnerships.

6. Operations Plan:

- **Production Methods:** Describe farming practices, including crop rotation, animal husbandry, and use of technology.
- **Supply Chain:** Outline logistics for sourcing inputs, managing inventory, and delivering products to customers.

7. Management and Organization:

- **Team Structure:** Define roles and responsibilities for farm management and operational staff.
- **Management Team:** Introduce key team members and their qualifications.

8. Financial Plan:

- **Startup Costs:** Estimate initial investment requirements, including land purchase, equipment, and infrastructure.
- **Revenue Projections:** Forecast sales revenue based on market demand and pricing strategies.

- **Budgets and Cash Flow:** Detail monthly or annual budgets, cash flow projections, and break-even analysis.

9. Risk Analysis and Mitigation:

- **Risk Assessment:** Identify potential risks and challenges, such as weather fluctuations, market volatility, or regulatory changes.
- **Risk Management Strategies:** Develop contingency plans and mitigation strategies to address identified risks.

10. Appendices:

- **Supporting Documents:** Include supplementary materials, such as resumes, legal documents, permits/licenses, and market research data.

Creating Your Farm Business Plan

To create a successful farm business plan, follow these steps:

1. Research and Gather Information:

Conduct thorough market research, assess your farm's resources, and gather relevant data to inform your plan.

2. Set Clear Goals and Objectives:

Define specific, measurable goals for your farm business, such as production targets, revenue milestones, and market share objectives.

3. Write and Structure Your Plan:

Organize your plan into sections, starting with an executive summary and following the components outlined above. Use clear, concise language and include supporting data and visuals where appropriate.

4. Review and Revise:

Seek feedback from trusted advisors, mentors, or industry experts. Revise your plan based on feedback to ensure clarity, accuracy, and feasibility.

5. Implement and Monitor:

Once finalized, use your business plan as a roadmap for launching and managing your farm business. Regularly review and update your plan to adapt to changing market conditions or business goals.

Conclusion

Developing a farm business plan is a critical step in laying the foundation for a successful and sustainable farm operation. By systematically outlining your farm's goals, strategies, and financial projections, you set yourself up for informed decision-making, efficient resource management, and growth opportunities. In the chapters ahead, we will delve deeper into executing your business plan, managing farm operations, and achieving your business objectives with confidence and clarity. Get ready to turn your farming aspirations into a well-defined roadmap for success!

Chapter 6: Securing Land for Your Farm

Securing suitable land is one of the foundational steps in starting a farm business. In this chapter, we'll explore the essential considerations, strategies, and steps involved in finding and acquiring the right piece of land for your farming venture.

Assessing Your Farming Needs

Before beginning your search for land, it's crucial to assess your farming needs and requirements. Consider the following factors:

1. Farm Enterprise:

Determine the type of farming enterprise you intend to pursue, whether it's crop farming, livestock raising, agroforestry, or a combination. Different enterprises have varying land requirements in terms of size, soil type, water access, and infrastructure.

2. Scale of Operations:

Decide on the scale of your farming operation—whether you plan to start small and expand gradually or launch a larger-scale operation from the outset. This decision will influence the size and features of the land you need.

3. Location and Climate:

Evaluate the climatic conditions suitable for your chosen crops or livestock. Consider factors such as temperature ranges, precipitation levels, frost dates, and microclimates within the region.

4. Infrastructure and Resources:

Assess the availability and adequacy of essential infrastructure such as water sources (wells, irrigation systems), access roads, fencing, storage facilities, and potential for future development (e.g., barns, greenhouses).

Steps to Finding and Acquiring Land

1. Research Local Real Estate Listings:

Start your search by exploring local real estate listings, land directories, and agricultural publications. Online platforms and local real estate agents specializing in rural properties can provide valuable insights and listings.

2. Network and Seek Local Connections:

Tap into local agricultural networks, farmer associations, and community organizations. Attend farmers' markets, agricultural fairs, and networking events to connect with landowners, farmers, and potential sellers.

3. Visit Potential Properties:

Schedule visits to prospective properties to assess their suitability for your farming needs. Consider factors such as soil quality, topography, drainage, sun exposure, and proximity to markets and suppliers.

4. Conduct Due Diligence:

Perform thorough due diligence on each property of interest. Verify zoning regulations, land use restrictions, water rights, environmental considerations (e.g., soil contamination), and any legal encumbrances.

5. Evaluate Financial Considerations:

Assess the financial aspects of land acquisition, including purchase price, property taxes, ongoing maintenance costs, and potential financing options (e.g., loans, grants, agricultural programs).

6. Negotiate and Secure Financing:

Negotiate terms with the seller or landowner, considering factors such as price adjustments, contingencies (e.g., soil testing), and financing arrangements. Secure financing or funding sources necessary for the purchase.

7. Finalize Legal and Contractual Obligations:

Consult with legal professionals experienced in agricultural real estate to finalize purchase agreements, contracts, and any legal documentation. Ensure all terms and conditions are clearly outlined and understood.

Sustainable Land Use Practices

1. Soil Health and Conservation:

Implement practices such as crop rotation, cover cropping, and organic farming methods to maintain soil fertility, structure, and health.

2. Water Management:

Utilize efficient irrigation systems, rainwater harvesting, and water conservation techniques to optimize water use and minimize environmental impact.

3. Biodiversity and Habitat Preservation:

Preserve natural habitats, plant hedgerows, and incorporate agroforestry practices to support biodiversity, pollinator habitats, and ecological resilience.

4. Energy Efficiency:

Adopt renewable energy sources (e.g., solar panels, wind turbines) and energy-efficient technologies to reduce greenhouse gas emissions and promote sustainable farming practices.

Conclusion

Securing land for your farm is a significant milestone that requires careful planning, research, and consideration of various factors. By assessing your farming needs, conducting thorough research, and navigating the acquisition process systematically, you can find the right piece of land to establish and grow your farm business. In the chapters ahead, we will explore strategies for land preparation, infrastructure development, and sustainable land management practices, ensuring you have the knowledge and tools to cultivate success in your farming journey. Get ready to transform your vision into reality on the fertile ground of your farm!

Chapter 7: Planning and Designing Your Farm Layout

Welcome to the exciting phase of planning and designing your farm layout! In this chapter, we'll explore the essential considerations, principles, and steps involved in creating a well-organized and efficient layout for your farm. Whether you're starting a new farm or reevaluating an existing one, careful planning and thoughtful design are key to optimizing productivity, sustainability, and overall success.

Importance of Farm Layout Design

A well-designed farm layout enhances operational efficiency, promotes optimal land use, and supports sustainable farming practices. It integrates infrastructure, facilitates workflow, and considers environmental factors, ensuring your farm operates smoothly and effectively. Here's why farm layout design is crucial:

1. Optimizing Resource Use:

A well-planned layout maximizes the use of available resources—land, water, sunlight, and labor—resulting in efficient production and reduced waste.

2. Enhancing Productivity:

By organizing farm components logically and strategically, you can streamline workflows, minimize operational bottlenecks, and increase overall productivity.

3. Promoting Sustainability:

Sustainable farm layout design incorporates practices that conserve natural resources, protect biodiversity, and minimize environmental impact.

4. Improving Safety and Accessibility:

Designing for safety includes proper placement of infrastructure, ergonomic considerations, and efficient movement of equipment and personnel.

5. Facilitating Expansion and Adaptation:

A flexible layout allows for future expansion, adaptation to changing market demands, and integration of new technologies or practices.

Steps to Planning Your Farm Layout

1. Define Your Objectives:

Clarify your farm goals, production targets, and long-term vision. Consider factors such as crop rotation, livestock management, and potential diversification.

2. Survey and Assess the Land:

Conduct a thorough survey of your farm site, noting topography, soil quality, drainage patterns, and existing natural features. This assessment informs layout decisions.

3. Identify Farm Zones and Functional Areas:

Divide your farm into zones based on the type of activities and infrastructure needed. Common zones include:

- **Production Area:** Main fields or areas for crop cultivation or livestock grazing.
- **Processing Area:** Facilities for washing, packing, and processing harvested products.
- **Storage Area:** Barns, sheds, or facilities for equipment storage, crop storage, and livestock housing.
- **Living Area:** Residential quarters or accommodations for farm workers, if applicable.

4. Plan Infrastructure and Facilities:

Designate locations for essential infrastructure and facilities, including:

- **Access Roads and Pathways:** Provide access to different farm zones while minimizing soil compaction and erosion.
- **Water Management Systems:** Implement irrigation systems, drainage channels, and water storage solutions to optimize water use and distribution.
- **Energy Sources:** Plan for energy needs, incorporating renewable energy options where feasible (e.g., solar panels, wind turbines).

5. Consider Environmental Factors:

Incorporate practices that enhance environmental sustainability and biodiversity, such as:

- **Natural Habitats:** Preserve and integrate natural habitats, hedgerows, or wildlife corridors.
- **Conservation Areas:** Designate areas for soil conservation practices, wetland protection, or native plant restoration.

6. Implement Efficient Workflow Patterns:

Plan for efficient movement of crops, livestock, and equipment. Minimize unnecessary travel distances and optimize workflow to reduce labor and energy costs.

7. Balance Aesthetic and Functional Design:

Aim for a layout that is both functional and visually appealing. Consider aesthetics in landscaping, building design, and integration of natural features.

Using Technology and Tools

1. Farm Design Software: Utilize farm design software or online tools to create detailed layouts, visualize plans, and optimize spatial arrangements.

2. Mapping and GIS: Use Geographic Information System (GIS) tools to analyze spatial data, map farm features, and make informed decisions based on land characteristics.

3. Precision Agriculture: Implement precision farming technologies such as GPS-guided equipment and sensor networks to monitor soil health, crop growth, and resource use efficiency.

Conclusion

Planning and designing your farm layout is a pivotal step in establishing a productive, efficient, and sustainable farm operation. By considering your farm's objectives, assessing land characteristics, and strategically organizing infrastructure and functional areas, you create a foundation for success in agriculture. In the chapters ahead, we will explore practical strategies for implementing your farm layout plan, managing operations effectively, and adapting to challenges and opportunities in the dynamic agricultural landscape. Get ready to transform your vision into a well-structured reality on the fertile ground of your farm!

Chapter 8: Choosing Crops and Livestock for Your Farm

Welcome to the chapter dedicated to choosing crops and livestock for your farm! This pivotal decision will shape the identity, productivity, and profitability of your farm business. Whether you're passionate about growing vegetables, raising animals, or both, selecting the right crops and livestock involves careful consideration of market demand, climate suitability, and personal preferences. Let's explore the essential factors and steps to help you make informed choices.

Understanding Your Market and Consumer Demand

Before deciding on which crops or livestock to raise, it's crucial to research and understand market demand. Consumer preferences, dietary trends, and local market dynamics play a significant role in determining the profitability and success of your farm products. Consider the following:

- **Market Trends:** Analyze current and projected trends in food consumption, health preferences, and sustainability practices.
- **Local Demand:** Assess the needs and preferences of your local community, including restaurants, farmers' markets, grocery stores, and direct consumers.
- **Value-Added Products:** Explore opportunities for value-added products such as organic produce, artisanal cheeses, or specialty meats that can command higher prices.

Factors to Consider When Choosing Crops

1. Climate and Soil Conditions:

Select crops that thrive in your local climate and soil type. Consider factors like temperature, precipitation, frost dates, and soil fertility.

2. Crop Rotation and Diversity:

Plan for crop rotation to maintain soil health and prevent pest and disease buildup. Diversify your crops to spread risk and optimize seasonal productivity.

3. Labor and Equipment Requirements:

Evaluate the labor intensity and equipment needed for each crop. Choose crops that align with your available resources and farming capabilities.

4. Marketability and Profitability:

Research crops with high market demand and profitability potential. Consider niche markets, specialty crops, or organic certification if suitable for your farm.

5. Personal Preference and Expertise:

Factor in your own interests, skills, and experience. Passion for a particular crop can enhance motivation and commitment to its success.

Factors to Consider When Choosing Livestock

1. Livestock Type and Purpose:

Determine whether you will raise animals for meat, dairy, fiber, eggs, or other products. Each type of livestock has specific requirements and management practices.

2. Breeds and Genetics:

Choose livestock breeds suited to your climate, production goals, and market preferences. Consider factors like growth rate, disease resistance, and temperament.

3. Housing and Infrastructure:

Plan for adequate housing, shelter, and fencing appropriate for the size and behavior of your livestock. Ensure facilities meet regulatory standards and animal welfare guidelines.

4. Feed and Nutrition:

Develop a feeding program based on nutritional requirements and availability of feed sources. Consider sustainable feeding practices and potential for pasture-based systems.

5. Health Management:

Implement a health management plan, including vaccination schedules, parasite control, and veterinary care. Maintain records to track animal health and performance.

Integrating Crops and Livestock (Mixed Farming)

1. Synergies and Benefits:

Explore integrated farming systems where crops and livestock complement each other. For example, livestock can provide manure for soil fertility, while crops can provide feed or forage.

2. Diversification and Risk Management:

Mixed farming diversifies income streams and spreads risk against market fluctuations, weather events, or disease outbreaks affecting either crops or livestock.

3. Resource Efficiency:

Optimize resource use through nutrient cycling, where waste products from one enterprise become inputs for another. This approach enhances sustainability and reduces environmental impact.

Planning for Success

1. Start Small and Expand:

Begin with a manageable scale and gradually expand based on experience, market demand, and available resources.

2. Market and Branding Strategy:

Develop a marketing plan to promote your farm products effectively. Consider branding, packaging, pricing strategies, and avenues for direct sales or partnerships.

3. Continuous Learning and Adaptation:

Stay informed about agricultural innovations, best practices, and market trends. Continuously adapt your crop and livestock choices based on performance feedback and changing consumer preferences.

Conclusion

Choosing the right crops and livestock for your farm is a critical decision that influences the success and sustainability of your agricultural enterprise. By considering market demand, climate suitability, personal interests, and integration opportunities, you can strategically select crops and livestock that align with your goals and resources. In the chapters ahead, we will delve deeper into farm management practices, production techniques, and strategies for optimizing crop and livestock performance. Get ready to cultivate your chosen crops and raise thriving livestock that contribute to the vibrant and diverse landscape of your farm!

Chapter 9: Farm Infrastructure and Equipment

Welcome to the chapter dedicated to farm infrastructure and equipment! Establishing the right infrastructure and acquiring appropriate equipment are crucial steps in setting up a functional and efficient farm operation. Whether you're starting a new farm or expanding an existing one, careful planning and investment in infrastructure and equipment can enhance productivity, optimize resource use, and support sustainable farming practices. Let's explore the essential considerations, components, and strategies for developing your farm infrastructure and selecting the right equipment.

Importance of Farm Infrastructure

Farm infrastructure refers to the physical facilities and structures necessary for farm operations, including buildings, utilities, and amenities. Well-designed infrastructure provides a foundation for efficiency, safety, and environmental stewardship on your farm. Here's why it's important:

- **Operational Efficiency:** Properly designed infrastructure facilitates smooth workflow, reduces labor inefficiencies, and enhances overall productivity.
- **Resource Management:** Infrastructure supports efficient use of resources such as water, energy, and land, promoting sustainability and cost-effectiveness.
- **Animal Welfare and Health:** Adequate infrastructure ensures comfortable housing, clean water, and appropriate ventilation for livestock, supporting their health and well-being.
- **Safety and Compliance:** Well-maintained infrastructure enhances farm safety, meets regulatory standards, and minimizes risks associated with accidents or environmental hazards.

Essential Components of Farm Infrastructure

1. Farm Buildings and Structures:

- **Barns and Shelters:** Provide housing and protection for livestock, equipment storage, and workspace for farm activities.
- **Processing Facilities:** Designated areas for washing, packing, and processing crops or livestock products.
- **Storage Facilities:** Barns or sheds for storing feed, equipment, tools, and harvested crops.
- **Greenhouses or High Tunnels:** Structures for extending the growing season, nurturing young plants, and protecting crops from adverse weather conditions.

2. Utilities and Services:

- **Water Supply:** Reliable water sources, irrigation systems, and watering points for crops and livestock.
- **Electricity and Lighting:** Access to electricity for lighting, equipment operation, and powering essential farm machinery.
- **Heating and Cooling:** Systems to regulate temperature in livestock housing, greenhouses, or processing areas.

3. Access and Mobility:

- **Access Roads and Pathways:** Well-maintained roads and pathways for vehicles, machinery, and personnel to navigate the farm safely and efficiently.
- **Fencing and Gates:** Perimeter fencing, paddocks, and gates for controlling livestock movement and protecting crops from wildlife.

4. Waste Management and Environmental Controls:

- **Composting Areas:** Facilities for composting organic waste materials to produce fertilizer.
- **Manure Management:** Systems for handling and utilizing livestock manure beneficially while minimizing environmental impact.
- **Environmental Controls:** Measures to mitigate soil erosion, manage runoff, and protect water quality.

Selecting Farm Equipment

Choosing the right equipment is essential for mechanizing farm tasks, reducing labor costs, and improving operational efficiency. Consider the following factors when selecting farm equipment:

1. Farm Size and Scale:

- Match equipment size and capacity to your farm's size and production scale. Choose equipment that can handle your workload efficiently without being underutilized or overburdened.

2. Type of Farming Operation:

- Different farming methods (e.g., conventional, organic) and enterprises (e.g., crop farming, livestock raising) require specific equipment. Consider equipment tailored to your farming practices and production goals.

3. Equipment Features and Capabilities:

- Evaluate features such as horsepower, fuel efficiency, reliability, and ease of maintenance. Opt for equipment with versatile attachments or implements to perform multiple tasks.

4. Budget and Financing:

- Determine your budget for equipment acquisition, including purchase costs, ongoing maintenance, and operational expenses. Explore financing options or leasing agreements if necessary.

5. Training and Support:

- Ensure access to training resources and technical support for equipment operation, maintenance, and troubleshooting.

Proper training enhances equipment efficiency and prolongs its lifespan.

Implementing Farm Infrastructure and Equipment

1. Planning and Design:

- Develop a detailed layout and design plan for farm infrastructure and equipment placement. Consider workflow patterns, accessibility, and future expansion needs.

2. Construction and Installation:

- Hire qualified contractors or utilize skilled labor to construct buildings, install utilities, and set up equipment according to design specifications and regulatory requirements.

3. Maintenance and Upkeep:

- Implement regular maintenance schedules for infrastructure and equipment to ensure functionality, prevent breakdowns, and prolong lifespan. Keep detailed records of maintenance activities.

4. Adaptation and Innovation:

- Stay informed about technological advancements and innovations in farm infrastructure and equipment. Evaluate opportunities to upgrade or integrate new technologies that enhance efficiency and sustainability.

Conclusion

Developing farm infrastructure and acquiring suitable equipment are critical investments that contribute to the success and sustainability of your farm business. By carefully planning, selecting appropriate components, and implementing efficient systems, you create a foundation for productive operations, animal welfare, environmental stewardship, and overall profitability. In the chapters ahead, we will explore strategies for managing farm operations, optimizing resource use, and adapting to evolving agricultural practices. Get ready to equip your farm with the tools and facilities needed to cultivate success in the dynamic world of agriculture!

Chapter 10: Farming Methods and Techniques

Welcome to the chapter dedicated to exploring various farming methods and techniques! As a farmer, choosing the right farming methods is crucial for maximizing productivity, sustainability, and profitability on your farm. Whether you're interested in traditional practices or innovative approaches like organic farming or agroecology, understanding different methods and techniques allows you to make informed decisions that align with your goals and values. Let's dive into the diverse world of farming methods and explore how they can benefit your agricultural enterprise.

Traditional vs. Modern Farming Methods

1. Traditional Farming Methods:

Traditional farming methods have been practiced for generations and often rely on conventional practices such as:

- **Conventional Tillage:** Plowing and tilling to prepare soil for planting.
- **Chemical Inputs:** Use of synthetic fertilizers, pesticides, and herbicides to manage pests and enhance crop yields.
- **Monoculture:** Growing single crops or raising one type of livestock on a larger scale.

While effective in certain contexts, traditional methods can pose challenges such as soil degradation, loss of biodiversity, and environmental impacts from chemical use.

2. Modern Farming Methods:

Modern farming embraces technological advancements and sustainable practices to improve efficiency and reduce environmental footprint:

- **Conservation Tillage:** Minimizing soil disturbance to improve soil health and reduce erosion.
- **Integrated Pest Management (IPM):** Combining biological, cultural, and mechanical methods to control pests and diseases.
- **Crop Rotation and Diversification:** Alternating crops to enhance soil fertility, manage pests naturally, and reduce reliance on chemicals.

Modern methods often emphasize sustainable agriculture principles, including resource efficiency, biodiversity conservation, and soil health improvement.

Sustainable Farming Methods

1. Organic Farming:

- **Principles:** Emphasizes soil health, biodiversity, and ecological balance without synthetic inputs. Uses natural fertilizers, biological pest control, and crop rotation
- **Certification:** Requires adherence to organic standards and certification processes to market products as organic.

2. Agroecology:

- **Approach:** Integrates ecological principles into farming systems, emphasizing biodiversity, soil health, and resilience.
- **Practices:** Includes agroforestry, polyculture, and integrating livestock to mimic natural ecosystems and enhance sustainability.

3. Permaculture:

- **Design Principles:** Uses design principles from natural ecosystems to create self-sustaining agricultural systems.
- **Practices:** Includes companion planting, water harvesting, and renewable energy integration to promote ecological harmony and resource efficiency.

Choosing the Right Farming Methods

1. Assess Your Goals and Values:

- Consider your farming objectives, values, and long-term sustainability goals when selecting farming methods.

2. Evaluate Environmental Conditions:

- Assess climate, soil type, water availability, and topography to determine which methods are most suitable for your farm.

3. Market Demand and Certification:

- Research consumer preferences and market demand for sustainably produced agricultural products. Consider organic certification or other eco-labels if aligned with your farming practices.

4. Start Small and Experiment:

- Begin with manageable scale and experiment with different methods on a smaller area of your farm. Monitor results and adapt practices based on performance and feasibility.

Implementing Farming Techniques

1. Soil Health Management:

- Conduct soil tests and implement practices such as cover cropping, composting, and mulching to improve soil fertility and structure.

2. Water Conservation and Management:

- Install efficient irrigation systems, practice rainwater harvesting, and implement water-saving techniques to optimize water use.

3. Integrated Pest and Disease Management:

- Monitor pests and diseases regularly. Implement biological controls, crop rotation, and resistant varieties to minimize chemical inputs.

4. Crop and Livestock Integration:

- Integrate livestock into cropping systems for nutrient cycling and pest control. Use rotational grazing and mixed farming practices to optimize land use.

Continuous Learning and Adaptation

1. Stay Informed:

- Keep abreast of new research, technologies, and best practices in sustainable agriculture through workshops, courses, and agricultural extension services.

2. Networking and Collaboration:

- Engage with other farmers, researchers, and organizations to exchange knowledge, share experiences, and collaborate on sustainable farming initiatives.

Conclusion

Choosing and implementing appropriate farming methods and techniques is fundamental to the success and sustainability of your farm business. By embracing sustainable practices, whether through organic farming, agroecology, or other innovative methods, you can enhance productivity, protect natural resources, and meet consumer demand for responsibly produced food. In the chapters ahead, we will explore practical strategies for farm management, marketing your products, and navigating challenges in the dynamic agricultural landscape. Get ready to cultivate success while nurturing the land and resources that sustain your farm!

Chapter 11: Farm Marketing and Sales Strategies

Welcome to the chapter dedicated to farm marketing and sales strategies! As a farmer, effectively marketing your products is essential for building a successful and sustainable farm business. Whether you're selling fresh produce, meats, dairy products, or value-added goods, understanding marketing principles and employing strategic sales tactics can help you reach your target audience, build customer loyalty, and increase profitability. Let's explore the key elements, strategies, and tips for effectively marketing your farm products.

Understanding Your Market

Before diving into marketing strategies, it's crucial to understand your target market and consumer preferences. Consider the following aspects:

- **Consumer Demographics:** Identify the demographics, preferences, and buying habits of your target customers (e.g., local families, health-conscious individuals, chefs at restaurants).
- **Market Trends:** Stay informed about current trends in food preferences, sustainability, and organic products to align your offerings with market demand.
- **Competitor Analysis:** Research other farms or businesses in your area offering similar products. Differentiate your farm by highlighting unique selling points (USPs) such as organic certification, sustainable practices, or local sourcing.

Developing Your Marketing Plan

1. Define Your Brand Identity:

- Establish a compelling brand identity that reflects your farm's values, mission, and unique story. This includes your farm name, logo, and brand messaging.

- Communicate what sets your farm apart—whether it's your commitment to organic practices, humane animal treatment, or heirloom varieties.

2. Product Offering and Pricing Strategy:

- Determine which products you will sell and their pricing strategy. Consider factors such as production costs, market prices, and perceived value by customers.
- Offer a variety of products to cater to different customer preferences and seasonal availability.

3. Sales Channels:

- Explore various sales channels to reach your target audience:
  - **Farmers' Markets:** Participate in local farmers' markets to connect directly with consumers, build relationships, and receive immediate feedback.
  - **Community Supported Agriculture (CSA):** Offer CSA subscriptions where customers receive regular deliveries of seasonal produce or products.
  - **Retail:** Partner with local grocery stores, co-ops, or specialty shops to sell your products.
  - **Online Sales:** Develop an e-commerce website or use online platforms to reach a wider audience beyond your local area.
  - **Farm Stands:** Set up on-farm stands or roadside markets for direct sales to passing customers.

4. Promotional Strategies:

- Utilize a mix of promotional tactics to raise awareness and drive sales:
  - **Social Media Marketing:** Use platforms like Instagram, Facebook, and Twitter to showcase farm activities, products, and engage with customers.

- **Content Marketing:** Share informative blogs, videos, or newsletters about farming practices, recipes, and seasonal updates to educate and attract customers.
- **Farm Tours and Events:** Host farm tours, workshops, or tasting events to engage the community, educate about your products, and build brand loyalty.
- **Collaborations and Partnerships:** Partner with local businesses, restaurants, or community organizations for joint promotions or events.

Building Customer Relationships

1. Customer Service Excellence:

- Provide exceptional customer service by being knowledgeable about your products, responsive to inquiries, and accommodating to customer preferences.
- Offer personalized experiences such as custom orders, recipe suggestions, or farm tours to enhance customer satisfaction.

2. Feedback and Improvement:

- Seek feedback from customers on product quality, packaging, and service. Use this input to make improvements and refine your offerings.
- Encourage customer reviews and testimonials to build trust and credibility among potential buyers.

Sustainability and Ethical Marketing

1. Transparent Communication:

- Transparently communicate your farming practices, sustainability initiatives, and ethical standards to build trust with consumers concerned about food origins and production methods.

- Highlight certifications, such as organic or fair-trade, to validate your commitment to sustainable farming practices.

2. Community Engagement:

- Engage with your local community through partnerships, educational initiatives, or charitable contributions. Demonstrating your farm's role in community support enhances your brand reputation.

Monitoring and Evaluation

1. Track Performance Metrics:

- Monitor sales data, customer feedback, and marketing analytics to evaluate the effectiveness of your strategies.
- Adjust your marketing plan based on insights gathered to optimize future campaigns and sales efforts.

2. Continuous Improvement:

- Stay adaptable and responsive to market changes, consumer preferences, and competitive dynamics. Continuously refine your marketing approach to stay relevant and competitive.

Conclusion

Effective marketing and sales strategies are essential for building a thriving farm business that resonates with consumers and contributes to your farm's sustainability. By understanding your market, developing a strong brand identity, utilizing diverse sales channels, and prioritizing customer relationships, you can successfully market your farm products and achieve long-term success. In the chapters ahead, we will explore additional aspects of farm management, production techniques, and navigating challenges in the agricultural industry. Get ready to grow your farm business while cultivating meaningful connections with your customers and community!

Chapter 12: Farm Financial Management and Planning

Welcome to the chapter dedicated to farm financial management and planning! As a farmer and business owner, understanding and effectively managing your finances are crucial for the long-term success and sustainability of your farm operation. Whether you're just starting out or looking to optimize your financial strategies, this chapter will guide you through essential principles, practices, and tools to ensure financial stability, growth, and profitability on your farm.

Importance of Financial Management

Sound financial management is the backbone of a successful farm business. It involves planning, monitoring, and controlling financial resources to achieve financial goals and sustain farm operations. Here's why it's essential:

- **Decision Making:** Helps in making informed decisions about investments, expansions, and operational changes.
- **Risk Management:** Identifies financial risks and implements strategies to mitigate them.
- **Long-term Sustainability:** Ensures profitability, cash flow management, and sustainable growth.

Essential Components of Farm Financial Management

1. Budgeting and Planning:

- **Annual Budget:** Develop a detailed budget that forecasts income and expenses for the upcoming year. Include production costs, labor, inputs, equipment maintenance, and marketing expenses.
- **Cash Flow Projection:** Forecast cash inflows and outflows to manage liquidity and ensure timely payment of bills and operational expenses.

2. Record Keeping and Accounting:

- **Financial Records:** Maintain accurate records of income, expenses, assets, and liabilities. Use accounting software or spreadsheets to track transactions and generate financial reports.
- **Financial Statements:** Prepare regular financial statements such as income statements, balance sheets, and cash flow statements to assess financial health and performance.

3. Financial Analysis:

- **Profitability Analysis:** Evaluate profitability of different enterprises, products, or markets. Calculate gross margins, net profits, and return on investment (ROI) to identify high-performing areas and opportunities for improvement.
- **Ratio Analysis:** Use financial ratios (e.g., liquidity ratios, debt-to-equity ratio) to assess financial health, solvency, and efficiency of farm operations.

4. Debt Management and Financing:

- **Debt Structure:** Manage existing debts and consider debt financing options for capital investments (e.g., land purchase, equipment upgrades).
- **Financing Options:** Explore financing options such as agricultural loans, grants, or government programs tailored to farmers.

5. Risk Management:

- **Insurance Coverage:** Evaluate insurance needs for property, crops, livestock, and liability. Insurance helps mitigate financial losses due to unexpected events such as natural disasters or crop failure.
- **Diversification:** Diversify income sources and production to spread risk and reduce dependence on a single enterprise or market.

Planning for Farm Expansion and Investment

1. Capital Investment Planning:

- Prioritize investments based on farm goals, profitability analysis, and long-term strategic plans.
- Consider investments in infrastructure improvements, technology upgrades, or expansion of productive capacity.

2. Succession Planning:

- Develop a succession plan to manage the transfer of farm ownership or management to the next generation. Address legal, financial, and operational aspects to ensure continuity and sustainability.

Financial Tools and Resources

1. **Financial Software:** Utilize accounting software (e.g., QuickBooks, FarmBooks) to streamline record-keeping, financial reporting, and budget management.

2. **Government Programs:** Explore agricultural grants, subsidies, or tax incentives available through government programs to support farm investments, conservation practices, or renewable energy projects.

3. **Financial Advisers:** Seek advice from agricultural financial consultants, accountants, or agricultural economists to develop customized financial strategies and improve financial decision-making.

Sustainable Financial Practices

1. **Resource Efficiency:** Implement practices that optimize resource use (e.g., water, energy) to reduce operating costs and improve profitability.

2. **Long-term Planning:** Develop long-term financial goals and strategies that align with sustainable farming practices, environmental stewardship, and community engagement.

Continuous Monitoring and Adaptation

1. **Regular Review:** Monitor financial performance regularly and compare actual results with budgeted projections. Identify variances and adjust strategies as needed.

2. **Adaptability:** Stay adaptable to changes in market conditions, regulatory environment, and agricultural trends. Continuously update financial plans and strategies to seize opportunities and mitigate risks.

Conclusion

Effective financial management is a cornerstone of farm success, ensuring profitability, resilience, and long-term sustainability. By implementing budgeting, record-keeping, financial analysis, and strategic planning practices, you can make informed decisions, optimize resources, and navigate financial challenges effectively. In the chapters ahead, we will delve deeper into farm operations, production techniques, and strategies for enhancing farm profitability and resilience in a dynamic agricultural landscape. Get ready to cultivate financial health and prosperity for your farm business!

Chapter 13: Farming Regulations and Compliance

Welcome to the chapter dedicated to farming regulations and compliance! As a farmer, navigating the regulatory landscape is essential to ensure legal compliance, environmental stewardship, and sustainable farm practices. Whether you're new to farming or expanding your operation, understanding agricultural regulations, permits, and best practices is crucial for operating a successful and responsible farm business. Let's explore the key regulations, compliance requirements, and strategies to navigate regulatory challenges effectively.

Understanding Agricultural Regulations

1. Local, State, and Federal Regulations:

- **Zoning and Land Use:** Familiarize yourself with local zoning laws and land use regulations that dictate where and how agricultural activities can be conducted.
- **Environmental Regulations:** Comply with federal and state environmental regulations governing water quality, air emissions, waste management, and conservation practices.
- **Food Safety Standards:** Adhere to food safety regulations such as Good Agricultural Practices (GAP), Hazard Analysis and Critical Control Points (HACCP), and FDA regulations for food handling and processing.

2. Permits and Licensing:

- Obtain necessary permits and licenses for farm operations, including water use permits, pesticide application licenses, animal welfare certifications, and business registrations.
- Stay updated on renewal requirements and regulatory changes affecting your farm activities.

3. Labor Laws and Employment Practices:

- Understand labor laws governing farm workers, including minimum wage, overtime regulations, worker safety (OSHA), and migrant worker protections (H-2A program).
- Implement fair labor practices, provide training on safety protocols, and maintain records to ensure compliance with employment regulations.

Environmental Stewardship and Conservation

1. Soil and Water Conservation:

- Implement soil conservation practices such as cover cropping, reduced tillage, and erosion control measures to protect soil health and minimize nutrient runoff.
- Comply with regulations related to wetland protection, water use efficiency, and nutrient management plans (NMPs) to safeguard water quality.

2. Wildlife and Habitat Protection:

- Adhere to regulations protecting wildlife habitats, endangered species, and migratory birds. Obtain permits for activities that may impact protected species or their habitats.
- Consider wildlife-friendly farming practices that promote biodiversity and habitat preservation alongside agricultural activities.

Food Safety and Quality Standards

1. Good Agricultural Practices (GAP):

- Implement GAP principles to minimize microbial contamination, ensure food safety, and maintain product quality from farm to consumer.
- Follow guidelines for handling, packaging, and transporting agricultural products to prevent contamination and meet market requirements.

2. Organic Certification:

- If pursuing organic farming, comply with USDA organic standards and undergo certification processes to label and market products as organic.
- Maintain records documenting organic practices, inputs used, and compliance with organic regulations to maintain certification.

Compliance Strategies and Best Practices

1. Stay Informed and Educated:

- Regularly monitor updates from regulatory agencies, agricultural extensions, and industry organizations to stay informed about changes in regulations and compliance requirements.
- Attend workshops, training sessions, or webinars on agricultural regulations and best practices to enhance your understanding and implementation.

2. Document Management:

- Maintain detailed records of farm activities, permits, inspections, and compliance documentation. Organize records for easy access and review during audits or inspections.
- Document practices related to soil conservation, pesticide applications, water usage, and waste management to demonstrate compliance with regulatory standards.

3. Proactive Communication:

- Establish positive relationships with regulatory agencies, local authorities, and community stakeholders. Seek guidance and clarification on regulations when needed.

- Communicate openly about your farm's practices, environmental initiatives, and commitment to regulatory compliance to build trust and transparency.

Challenges and Resources

1. Navigating Complexity:

- Recognize challenges associated with regulatory compliance, including administrative burdens, costs of compliance, and interpretation of ambiguous regulations.
- Utilize resources such as agricultural advisors, legal counsel, or industry associations to navigate complex regulatory issues and ensure compliance.

2. Continuous Improvement:

- Embrace a culture of continuous improvement by evaluating and updating your farm practices in response to regulatory changes, technological advancements, and evolving best practices.
- Engage in sustainability initiatives, voluntary conservation programs, or certifications that demonstrate your commitment to environmental stewardship and responsible farming practices.

Conclusion

Navigating agricultural regulations and compliance requirements is essential for operating a successful and sustainable farm business. By understanding local, state, and federal regulations, implementing best practices for environmental stewardship, and maintaining documentation of farm activities, you can ensure legal compliance, protect natural resources, and uphold food safety standards. In the chapters ahead, we will explore additional aspects of farm management, production techniques, and strategies for optimizing farm operations in a regulated environment. Get ready to cultivate compliance while growing your farm business responsibly and ethically!

Chapter 14: Farm Risk Management

Welcome to the chapter dedicated to farm risk management! As a farmer, understanding and mitigating risks are crucial to protect your farm business from potential threats and uncertainties. Agricultural operations are inherently exposed to various risks, including market volatility, weather events, pests, and regulatory changes. Effective risk management strategies not only minimize potential losses but also enhance resilience, profitability, and long-term sustainability. In this chapter, we will explore key risks faced by farmers and practical strategies to manage and mitigate these risks effectively.

Identifying Farm Risks

1. Production Risks:

- **Weather Events:** Adverse weather conditions such as droughts, floods, frosts, and storms can affect crop yields, livestock health, and farm operations.
- **Pests and Diseases:** Outbreaks of pests, insects, and diseases can damage crops, reduce yields, and impact farm profitability.
- **Crop and Livestock Management:** Challenges related to crop failure, livestock health issues, or production inefficiencies due to poor management practices.

2. Market Risks:

- **Price Volatility:** Fluctuations in commodity prices and market demand can affect farm revenue and profitability.
- **Market Access:** Dependence on a limited number of buyers or markets, changes in consumer preferences, and competition from imported products.
- **Marketing and Sales:** Risks related to product pricing, marketing strategies, and distribution channels.

3. Financial Risks:

- **Debt and Financing:** Risks associated with loan repayments, interest rates, and financial obligations.
- **Cash Flow Management:** Challenges in managing cash flow, meeting financial obligations, and covering operational expenses during lean periods.
- **Income Variability:** Uncertainty in income due to seasonal fluctuations, market conditions, or unexpected expenses.

4. Legal and Regulatory Risks:

- **Compliance Issues:** Risks related to regulatory changes, environmental regulations, permits, and legal liabilities.
- **Contractual Obligations:** Risks associated with contracts, agreements, and disputes with suppliers, buyers, or business partners.

Strategies for Farm Risk Management

1. Risk Assessment and Planning:

- Conduct a comprehensive risk assessment to identify potential threats and vulnerabilities specific to your farm operation.
- Prioritize risks based on their impact and likelihood, considering both short-term and long-term implications.

2. Diversification:

- **Crop and Livestock Diversification:** Planting a variety of crops or raising multiple livestock species can spread production risks and reduce dependence on a single enterprise.
- **Income Diversification:** Explore alternative revenue streams such as agritourism, value-added products, or direct-to-consumer sales (e.g., farmers' markets, CSA subscriptions).

3. Insurance and Risk Transfer:

- **Crop Insurance:** Purchase crop insurance policies to protect against losses due to adverse weather, pests, or yield reductions.
- **Livestock Insurance:** Obtain livestock insurance coverage for protection against disease outbreaks, accidents, or mortality.
- **Business Interruption Insurance:** Consider insurance policies that cover losses from market disruptions, natural disasters, or other unforeseen events.

4. Financial Management:

- **Budgeting and Cash Flow Planning:** Develop and maintain a realistic budget that accounts for production costs, operating expenses, and contingencies.
- **Emergency Fund:** Build an emergency fund to cover unexpected expenses or income shortfalls during challenging periods.

5. Sustainable Practices:

- **Soil Health and Conservation:** Implement sustainable agriculture practices such as cover cropping, conservation tillage, and nutrient management to enhance soil health and resilience to weather extremes.
- **Water Management:** Invest in efficient irrigation systems, rainwater harvesting, and water conservation practices to mitigate risks associated with water scarcity or drought.

6. Monitoring and Adaptation:

- **Regular Monitoring:** Monitor key performance indicators (KPIs), market trends, and weather forecasts to anticipate and respond to potential risks proactively.
- **Adaptive Strategies:** Remain flexible and adaptive in adjusting production plans, marketing strategies, and operational decisions based on changing conditions and emerging risks.

Collaborative Approaches and Resources

1. **Industry Networks and Associations:**

   - Engage with agricultural networks, cooperative extensions, and industry associations for access to resources, information, and collaborative risk management initiatives.
   - Participate in knowledge-sharing forums, workshops, and peer-learning opportunities to exchange ideas and best practices with other farmers.

2. **Government Programs and Support:**

   - Explore government-supported risk management programs, grants, or subsidies that provide financial assistance, technical support, or risk mitigation tools for farmers.
   - Stay informed about disaster assistance programs, emergency relief funds, and agricultural policies that support resilience and recovery efforts.

Conclusion

Effective farm risk management is essential for safeguarding your farm business, ensuring resilience in the face of challenges, and maximizing long-term sustainability. By identifying risks, implementing proactive strategies, and leveraging resources, you can minimize potential losses, seize opportunities, and navigate uncertainties in the dynamic agricultural environment. In the chapters ahead, we will delve deeper into farm operations, sustainable practices, and strategies for enhancing farm resilience and profitability. Get ready to cultivate resilience and thrive amidst agricultural challenges with informed risk management!

Chapter 15: Sustainable Agriculture Practices

Welcome to the chapter dedicated to sustainable agriculture practices! In today's agricultural landscape, sustainability is more than just a buzzword—it's a fundamental approach to farming that emphasizes environmental stewardship, economic viability, and social responsibility. Sustainable agriculture aims to meet the needs of the present without compromising the ability of future generations to meet their own needs. Whether you're a small-scale farmer or managing a larger operation, adopting sustainable practices not only preserves natural resources but also enhances farm productivity, resilience, and profitability. Let's explore the principles, benefits, and practical strategies for integrating sustainable agriculture into your farm.

Principles of Sustainable Agriculture

1. Environmental Stewardship:

- **Soil Health:** Implement practices such as cover cropping, crop rotation, and reduced tillage to improve soil fertility, structure, and biodiversity.
- **Water Conservation:** Use efficient irrigation systems, rainwater harvesting, and water-saving techniques to minimize water usage and protect water quality.
- **Biodiversity:** Promote biodiversity by planting hedgerows, cover crops, and native plants to support beneficial insects, pollinators, and wildlife habitats.

2. Economic Viability:

- **Resource Efficiency:** Optimize resource use (e.g., inputs, energy) to reduce costs, improve efficiency, and enhance profitability over the long term.
- **Diversification:** Explore diversified income streams, value-added products, and direct marketing channels (e.g., farmers' markets, CSA subscriptions) to stabilize farm income and mitigate market risks.

- **Long-term Planning:** Develop strategic plans that prioritize investments in sustainable infrastructure, technology, and practices that yield economic returns while benefiting the environment.

3. Social Responsibility:

- **Community Engagement:** Engage with local communities through education programs, farm tours, and partnerships to foster understanding and support for sustainable agriculture.
- **Fair Labor Practices:** Ensure fair wages, safe working conditions, and respect for farm workers' rights to promote social equity and well-being within the farming community.
- **Consumer Education:** Educate consumers about the benefits of sustainable farming practices, transparency in food production, and the importance of supporting local, sustainable agriculture.

Benefits of Sustainable Agriculture

1. Environmental Benefits:

- **Climate Resilience:** Reduce greenhouse gas emissions, mitigate climate change impacts, and build resilience to extreme weather events through sustainable practices.
- **Soil Conservation:** Preserve soil health, prevent erosion, and maintain nutrient cycling to sustain long-term productivity and ecosystem services.
- **Water Quality:** Protect water resources from contamination, minimize runoff, and enhance water-use efficiency through conservation practices.

2. Economic Benefits:

- **Cost Savings:** Lower input costs, reduced reliance on external inputs (e.g., fertilizers, pesticides), and improved operational efficiency contribute to cost savings and profitability.

- **Market Access:** Meet growing consumer demand for sustainably produced food and agricultural products, potentially commanding premium prices and accessing niche markets.
- **Resilience to Market Volatility:** Diversified income streams and reduced dependency on external inputs buffer against market fluctuations and economic uncertainties.

3. Social Benefits:

- **Community Resilience:** Strengthen local economies, create job opportunities, and support rural development through sustainable agricultural practices.
- **Health and Well-being:** Provide nutritious, safe food products while enhancing public health outcomes and promoting well-being in communities.

Practical Strategies for Sustainable Farming

1. Crop and Soil Management:

- Implement crop rotations, cover cropping, and integrated pest management (IPM) to enhance soil fertility, suppress pests naturally, and reduce reliance on synthetic inputs.
- Adopt no-till or reduced tillage practices to improve soil structure, retain moisture, and sequester carbon in the soil.

2. Water Management:

- Install efficient irrigation systems (e.g., drip irrigation, micro-sprinklers) to deliver water directly to plant roots, minimize evaporation, and optimize water use efficiency.
- Employ water-saving techniques such as mulching, rainwater harvesting, and soil moisture monitoring to conserve water resources.

3. Energy Efficiency:

- Invest in renewable energy sources (e.g., solar panels, wind turbines) to reduce reliance on fossil fuels, lower energy costs, and minimize carbon footprint.
- Implement energy-efficient practices in farm operations, equipment use, and building design to maximize energy savings and sustainability.

4. Livestock and Animal Management:

- Practice rotational grazing, managed intensive grazing, or silvopasture to improve pasture health, soil fertility, and carbon sequestration.
- Use humane animal husbandry practices, provide access to natural habitats, and prioritize animal welfare to enhance productivity and ethical standards.

Monitoring and Continuous Improvement

1. Performance Metrics:

- Monitor key performance indicators (KPIs) such as soil health indicators, water use efficiency, energy consumption, and biodiversity metrics to assess progress toward sustainability goals.
- Use data-driven insights to identify areas for improvement, adjust management practices, and optimize resource allocation.

2. Education and Innovation:

- Stay informed about advancements in sustainable agriculture, participate in educational workshops, and collaborate with agricultural advisors, researchers, and industry experts.
- Embrace innovation in technology, research new practices, and pilot projects that enhance sustainability, resilience, and productivity on your farm.

Conclusion

Sustainable agriculture is a holistic approach that balances environmental stewardship, economic viability, and social responsibility to ensure the long-term health and resilience of farming systems. By adopting sustainable practices, you can protect natural resources, enhance farm profitability, and contribute to healthier communities and ecosystems. In the chapters ahead, we will delve deeper into farm management strategies, production techniques, and innovative solutions for advancing sustainable agriculture in a changing world. Get ready to cultivate sustainability and thrive as a steward of the land and advocate for sustainable farming practices!

Chapter 16: Farm Technology and Innovation

Welcome to the chapter dedicated to farm technology and innovation! In today's rapidly evolving agricultural landscape, technology plays a pivotal role in enhancing efficiency, productivity, and sustainability on farms of all sizes. From precision agriculture to advanced machinery and digital tools, embracing technology empowers farmers to make data-driven decisions, optimize resource use, and overcome challenges posed by climate change and global food demand. This chapter explores the transformative impact of technology on modern farming practices, key innovations, and strategies for integrating technology into your farm operation.

The Role of Technology in Modern Farming

1. Precision Agriculture:

- **Remote Sensing and GIS:** Use satellite imagery, drones, and Geographic Information Systems (GIS) to monitor crop health, soil conditions, and field variability. This data enables targeted interventions for precise application of inputs such as fertilizers and pesticides.
- **Precision Planting and Irrigation:** Employ GPS-guided equipment for precise planting, seeding, and irrigation management. Adjust seeding rates and water application based on real-time field data to optimize crop growth and resource efficiency.

2. Digital Farm Management Tools:

- **Farm Management Software:** Utilize integrated software platforms for planning, record-keeping, and analysis of farm operations. Track expenses, manage inventory, and generate financial reports to streamline administrative tasks and improve decision-making.
- **Weather Forecasting and Monitoring:** Access weather forecasts and real-time weather data through mobile apps or online

platforms. Plan field activities, manage irrigation schedules, and mitigate risks associated with weather variability.

3. Smart Farm Equipment and Robotics:

- **Autonomous Vehicles:** Deploy autonomous tractors, harvesters, and drones for tasks such as planting, spraying, and crop monitoring. Enhance operational efficiency, reduce labor costs, and minimize environmental impact.
- **Robotics and AI:** Integrate robotic systems for precision weeding, harvesting, and sorting of crops. AI algorithms analyze data to optimize farming practices, improve yield predictions, and detect anomalies in crop health.

Advantages of Adopting Farm Technology

1. Enhanced Efficiency and Productivity:

- Automate repetitive tasks, reduce manual labor, and optimize workflows with technology-driven solutions. Increase farm output while minimizing input costs and operational inefficiencies.
- Improve timeliness of operations such as planting, harvesting, and pest management, leading to higher yields and better crop quality.

2. Resource Conservation:

- **Water Management:** Implement precision irrigation systems that deliver water directly to crops based on soil moisture levels and weather conditions. Reduce water usage, minimize runoff, and improve water-use efficiency.
- **Energy Efficiency:** Use energy-efficient equipment, renewable energy sources, and smart grid technologies to reduce carbon footprint and lower operational costs.

3. Data-Driven Decision Making:

- Collect and analyze farm data (e.g., yield maps, soil samples, weather patterns) to gain insights into crop performance, soil health, and operational efficiencies.
- Make informed decisions on crop selection, input application, pest management strategies, and resource allocation to optimize yields and profitability.

Integrating Technology into Farm Operations

1. Assessing Farm Needs and Goals:

- Identify specific challenges, goals, and areas where technology can address operational inefficiencies or enhance productivity.
- Conduct a technology readiness assessment and consider factors such as cost, compatibility with existing systems, and potential return on investment (ROI).

2. Training and Adoption:

- Provide training and education to farm operators and staff on using new technologies effectively. Ensure proficiency in equipment operation, data interpretation, and troubleshooting.
- Foster a culture of innovation and continuous learning to embrace technological advancements and adapt to evolving agricultural practices.

3. Collaboration and Networking:

- Engage with agricultural research institutions, extension services, and technology providers to stay informed about emerging technologies, research findings, and best practices.
- Participate in industry conferences, field demonstrations, and peer-learning networks to exchange knowledge, share experiences, and collaborate on innovative solutions.

Overcoming Challenges and Considerations

1. Cost and Investment:

- Evaluate upfront costs, ongoing maintenance expenses, and potential financial risks associated with adopting new technologies.
- Explore financing options, grants, or government programs that support technology adoption and modernization efforts in agriculture.

2. Data Security and Privacy:

- Implement robust cybersecurity measures to protect sensitive farm data, intellectual property, and customer information stored on digital platforms.
- Adhere to data privacy regulations and best practices for data management, storage, and sharing with third-party service providers.

3. Sustainability and Ethical Considerations:

- Ensure that technological advancements align with principles of sustainable agriculture, environmental stewardship, and ethical farming practices.
- Monitor and mitigate potential environmental impacts, such as increased energy consumption or disruption of natural ecosystems, associated with intensive technology use.

Looking Ahead: Future Trends in Farm Technology

1. Blockchain and Traceability: Utilize blockchain technology for enhanced traceability of food products, supply chain transparency, and verification of sustainability claims.

2. Artificial Intelligence (AI) and Predictive Analytics: Harness AI-powered analytics to predict crop yields, optimize resource allocation, and mitigate risks from pests, diseases, and weather extremes.

3. Climate Smart Agriculture: Develop and adopt technologies that support climate adaptation and mitigation strategies, such as carbon sequestration, soil carbon monitoring, and resilient crop varieties.

Conclusion

Farm technology and innovation are revolutionizing agriculture, offering unprecedented opportunities to enhance efficiency, sustainability, and profitability. By integrating advanced technologies, leveraging data-driven insights, and embracing continuous innovation, farmers can navigate challenges, seize opportunities, and build resilient farming systems for the future. In the chapters ahead, we will explore additional aspects of farm management, sustainable practices, and strategies for thriving in a dynamic agricultural landscape. Get ready to cultivate innovation and harness the power of technology to transform your farm operation!

Chapter 17: Marketing and Branding Your Farm Products

Welcome to the chapter dedicated to marketing and branding your farm products! In today's competitive market, how you present and sell your products can be as important as the quality of the products themselves. Effective marketing and strong branding not only attract customers but also build loyalty and trust, ensuring repeat business and word-of-mouth referrals. This chapter will guide you through the essentials of creating a compelling brand, developing a marketing strategy, and leveraging various channels to reach your target audience.

Understanding Your Market

1. Market Research:

- **Identify Your Target Audience:** Understand who your customers are. Are they health-conscious consumers, local chefs, grocery stores, or families looking for fresh produce? Knowing your audience helps tailor your marketing efforts effectively.
- **Analyze Market Trends:** Stay updated on industry trends, consumer preferences, and market demands. This helps you adapt and offer products that are in demand.
- **Competitive Analysis:** Study your competitors to understand their strengths and weaknesses. Identify gaps in the market that your farm can fill.

2. Defining Your Unique Selling Proposition (USP):

- **Highlight What Makes You Unique:** Whether it's organic certification, sustainable farming practices, heirloom varieties, or exceptional taste, your USP sets you apart from the competition.
- **Communicate Your Values:** Share your farm's story, values, and mission. Consumers are increasingly interested in the story behind their food and the values of the businesses they support.

Building Your Brand

1. Creating a Strong Brand Identity:

- **Logo and Visual Identity:** Design a memorable logo and cohesive visual elements (colors, fonts, imagery) that reflect your farm's personality and values.
- **Brand Voice:** Develop a consistent tone and style for your communications, whether it's friendly and approachable, professional and informative, or passionate and enthusiastic.

2. Crafting Your Brand Message:

- **Mission Statement:** Clearly articulate your farm's mission and vision. This forms the foundation of your brand message and helps build a connection with your audience.
- **Tagline:** Create a catchy and memorable tagline that encapsulates your brand's essence. It should be short, impactful, and easy to remember.

Developing a Marketing Strategy

1. Setting Marketing Goals:

- **Define Clear Objectives:** Set specific, measurable, achievable, relevant, and time-bound (SMART) goals for your marketing efforts. Examples include increasing sales by 20%, expanding customer base, or boosting social media engagement.
- **Budgeting:** Allocate a budget for marketing activities. This includes costs for advertising, promotions, packaging, and other marketing materials.

2. Choosing Marketing Channels:

- **Direct Marketing:** Sell your products directly to consumers through farmers' markets, CSA (Community Supported Agriculture) programs, farm stands, and online stores.

- **Wholesale and Retail:** Establish relationships with local grocery stores, restaurants, and co-ops to supply them with your products.
- **Digital Marketing:** Leverage social media, email marketing, and a farm website to reach a broader audience and engage with customers online.

Leveraging Digital Marketing

1. Social Media:

- **Platforms:** Choose platforms where your target audience is most active, such as Facebook, Instagram, Twitter, and Pinterest.
- **Content Creation:** Share engaging content, including photos, videos, farm updates, recipes, and behind-the-scenes stories. Highlight your products, farming practices, and customer testimonials.
- **Engagement:** Interact with your followers by responding to comments, messages, and reviews. Host giveaways, contests, and live Q&A sessions to foster community engagement.

2. Email Marketing:

- **Build an Email List:** Collect email addresses at farmers' markets, through your website, and at events. Offer incentives such as discounts or free recipes for sign-ups.
- **Newsletters:** Send regular newsletters with farm updates, product availability, seasonal recipes, and special offers. Keep the content valuable and engaging to encourage open rates and clicks.

3. Website and E-commerce:

- **User-Friendly Website:** Ensure your website is easy to navigate, mobile-friendly, and provides essential information about your farm, products, and purchasing options.

- **Online Store:** Set up an e-commerce platform to sell your products directly to consumers. Offer convenient payment options and reliable shipping or pick-up services.

Traditional Marketing Methods

1. Print Materials:

- **Brochures and Flyers:** Distribute brochures and flyers at local events, farmers' markets, and retail locations. Include information about your farm, products, and contact details.
- **Business Cards:** Always have business cards on hand to share with potential customers, partners, and suppliers.

2. Community Engagement:

- **Farm Tours and Events:** Host farm tours, workshops, and events to invite the community to experience your farm firsthand. This builds trust and strengthens relationships with customers.
- **Local Partnerships:** Collaborate with local businesses, schools, and organizations to promote your farm and products. Partnerships can include joint events, sponsorships, and cross-promotions.

Measuring and Adjusting Your Strategy

1. Tracking Performance:

- **Analytics:** Use tools like Google Analytics, social media insights, and email marketing metrics to track the performance of your marketing efforts.
- **Customer Feedback:** Collect feedback from customers through surveys, reviews, and direct interactions. Understand their preferences and areas for improvement.

2. Adjusting Strategies:

- **Analyze Results:** Regularly review your marketing performance against your goals. Identify what's working well and what needs improvement.
- **Flexibility:** Be willing to adapt and try new approaches based on the insights you gather. Marketing is an ongoing process that requires flexibility and creativity.

Conclusion

Effective marketing and branding are essential for the success of your farm business. By understanding your market, building a strong brand, and leveraging various marketing channels, you can attract and retain loyal customers. Remember, the key to successful marketing is authenticity—stay true to your farm's values and mission, and let your passion for farming shine through in every interaction. In the chapters ahead, we will continue to explore strategies for optimizing farm operations, sustainable practices, and innovative solutions to ensure your farm thrives in a dynamic agricultural landscape. Get ready to cultivate connections and watch your farm business flourish!

Chapter 18: Financial Management for Your Farm

Welcome to the chapter on financial management for your farm! Running a successful farm business requires more than just growing crops or raising livestock; it involves meticulous financial planning, budgeting, and monitoring to ensure long-term sustainability and profitability. This chapter will guide you through essential financial management practices, from understanding farm finances to creating budgets and managing cash flow. Let's delve into how you can effectively manage your farm's finances to achieve your business goals.

Understanding Farm Finances

1. Key Financial Statements:

- **Income Statement:** This statement provides a summary of your farm's revenues, expenses, and profits over a specific period. It helps you understand your farm's profitability.
- **Balance Sheet:** This statement lists your farm's assets, liabilities, and owner's equity at a given point in time. It provides a snapshot of your farm's financial health.
- **Cash Flow Statement:** This statement tracks the flow of cash in and out of your farm. It helps you manage liquidity and ensure you have enough cash to cover expenses.

2. Revenue Sources:

- **Product Sales:** Income from selling crops, livestock, dairy, or other farm products.
- **Subsidies and Grants:** Financial support from government programs, agricultural organizations, or grants.
- **Diversified Income:** Additional revenue from agritourism, farm stays, workshops, or value-added products.

Creating a Farm Budget

1. Setting Financial Goals:

- **Short-term Goals:** These might include purchasing new equipment, increasing production, or expanding market reach within the next year.
- **Long-term Goals:** These could involve land acquisition, infrastructure development, or transitioning to organic farming over several years.

2. Budget Categories:

- **Income:** Estimate all potential income sources for the year.
- **Fixed Costs:** Include expenses that remain constant regardless of production levels, such as property taxes, insurance, and loan payments.
- **Variable Costs:** These are expenses that fluctuate with production levels, such as seeds, feed, fuel, and labor.
- **Capital Expenditures:** Budget for significant investments like machinery, buildings, or irrigation systems.

3. Budgeting Tips:

- **Be Realistic:** Base your budget on realistic estimates of income and expenses, considering historical data and market trends.
- **Contingency Planning:** Set aside a portion of your budget for unexpected expenses or emergencies.
- **Review and Adjust:** Regularly review your budget and adjust as needed based on actual performance and changing circumstances.

Managing Cash Flow

1. Cash Flow Cycles:

- Understand your farm's cash flow cycles, which might vary seasonally depending on planting and harvest schedules or livestock breeding periods.

2. Improving Cash Flow:

- **Sales Planning:** Time your product sales to align with periods of high demand and favorable market prices.
- **Payment Terms:** Negotiate favorable payment terms with suppliers and customers to maintain positive cash flow.
- **Inventory Management:** Manage inventory efficiently to avoid overproduction and reduce storage costs.

Financing Your Farm

1. Types of Financing:

- **Loans:** Obtain loans from banks, credit unions, or agricultural lenders for capital investments, operating expenses, or expansion projects.
- **Grants and Subsidies:** Apply for grants and subsidies from government programs or agricultural organizations to support specific initiatives or innovations.
- **Equity Financing:** Consider partnerships, investors, or crowdfunding to raise capital without taking on debt.

2. Choosing the Right Financing:

- **Assess Needs:** Determine the amount of funding required, the purpose of the financing, and the repayment terms.
- **Evaluate Options:** Compare different financing options based on interest rates, repayment terms, eligibility criteria, and potential impact on your business.
- **Plan Repayment:** Develop a clear repayment plan that aligns with your farm's cash flow and financial projections.

Monitoring and Adjusting Financial Plans

1. Regular Financial Reviews:

- Conduct regular reviews of your financial statements, budgets, and cash flow to assess performance against goals and identify areas for improvement.

- **Monthly Reviews:** Monitor monthly income and expenses to ensure you're on track with your budget.
- **Quarterly and Annual Reviews:** Evaluate broader financial performance and make strategic adjustments based on trends and outcomes.

2. Adjusting Strategies:

- **Expense Management:** Identify areas where you can reduce costs without compromising quality or productivity.
- **Income Enhancement:** Explore opportunities to increase revenue through new markets, products, or value-added services.
- **Risk Management:** Implement strategies to mitigate financial risks, such as diversifying income sources, purchasing insurance, or building emergency funds.

Leveraging Technology for Financial Management

1. Farm Management Software:

- Use farm management software to streamline financial record-keeping, budgeting, and reporting. These tools can help automate processes, reduce errors, and provide valuable insights.

2. Mobile Apps and Tools:

- Utilize mobile apps for on-the-go expense tracking, invoicing, and cash flow monitoring. These tools can help you stay organized and make informed decisions in real-time.

Building Financial Resilience

1. Emergency Fund:

- Establish an emergency fund to cover unexpected expenses, such as equipment repairs, natural disasters, or market downturns. This fund acts as a financial safety net.

2. Diversification:

- Diversify your farm's income streams to reduce dependency on a single source and enhance financial stability. Consider adding agritourism, educational programs, or value-added products to your portfolio.

Conclusion

Effective financial management is the backbone of a successful farm business. By understanding your finances, creating realistic budgets, managing cash flow, and leveraging financing options, you can ensure the long-term sustainability and profitability of your farm. Remember, financial management is an ongoing process that requires regular monitoring, adjustments, and strategic planning. In the chapters ahead, we will explore additional aspects of farm operations, innovative solutions, and sustainable practices to help you thrive in the dynamic agricultural landscape. Get ready to cultivate financial acumen and secure the future of your farm business!

Chapter 19: Sustainable Farming Practices

Welcome to the chapter on sustainable farming practices! In an era where environmental concerns are paramount, adopting sustainable practices is not only a moral responsibility but also a smart business strategy. Sustainable farming ensures that your farm remains productive for generations to come while preserving the environment, enhancing biodiversity, and contributing to the well-being of your community. This chapter will delve into various sustainable farming practices, their benefits, and how to implement them on your farm.

Understanding Sustainable Farming

1. Definition and Principles:

- **Sustainability:** Sustainable farming aims to meet current agricultural needs without compromising the ability of future generations to meet their own needs. It encompasses environmental health, economic profitability, and social equity.
- **Core Principles:** These include maintaining healthy soil, conserving water, reducing pollution, promoting biodiversity, and ensuring the well-being of farm workers and local communities.

Soil Health Management

1. Soil Testing and Amendments:

- **Regular Soil Testing:** Conduct soil tests to understand the nutrient composition and pH levels of your soil. This helps in making informed decisions about fertilization and amendments.
- **Organic Amendments:** Use compost, manure, and cover crops to improve soil structure, increase organic matter, and enhance microbial activity.

2. Crop Rotation and Diversity:

- **Crop Rotation:** Practice rotating crops to break pest and disease cycles, improve soil fertility, and reduce soil erosion.
- **Intercropping and Polyculture:** Grow multiple crops together to promote biodiversity, optimize space, and improve soil health.

Water Conservation

1. Efficient Irrigation Systems:

- **Drip Irrigation:** Install drip irrigation systems to deliver water directly to the plant roots, reducing water waste and evaporation.
- **Rainwater Harvesting:** Collect and store rainwater for irrigation purposes. This reduces dependency on groundwater and municipal water sources.

2. Soil Moisture Management:

- **Mulching:** Apply organic or inorganic mulch to the soil surface to retain moisture, suppress weeds, and regulate soil temperature.
- **Soil Organic Matter:** Increase soil organic matter to improve water retention and reduce irrigation needs.

Integrated Pest Management (IPM)

1. Biological Controls:

- **Beneficial Insects:** Introduce beneficial insects like ladybugs, lacewings, and predatory mites to control pest populations naturally.
- **Natural Predators:** Encourage natural predators such as birds and bats by providing habitats like nesting boxes and hedgerows.

2. Cultural Practices:

- **Crop Rotation:** Rotate crops to disrupt pest life cycles and reduce pest buildup.
- **Sanitation:** Keep fields clean of crop residues and debris that can harbor pests.

Renewable Energy and Resource Efficiency

1. Renewable Energy Sources:

- **Solar Power:** Install solar panels to generate electricity for farm operations, reducing reliance on fossil fuels and lowering energy costs.
- **Wind Power:** Utilize wind turbines in areas with consistent wind patterns to generate renewable energy for your farm.

2. Energy Efficiency:

- **Energy-Efficient Equipment:** Invest in energy-efficient machinery and equipment to reduce energy consumption and lower operational costs.
- **Farm Buildings:** Improve the insulation and energy efficiency of farm buildings to reduce heating and cooling needs.

Promoting Biodiversity

1. Habitat Creation:

- **Wildlife Habitats:** Create habitats such as hedgerows, wetlands, and buffer strips to support a diverse range of wildlife.
- **Pollinator Gardens:** Plant flowering plants and maintain natural areas to attract and support pollinators like bees, butterflies, and hummingbirds.

2. Agroforestry:

- **Silvopasture:** Integrate trees and shrubs into pastureland to provide shade for livestock, improve soil health, and enhance biodiversity.
- **Alley Cropping:** Grow crops between rows of trees to diversify production, improve soil health, and provide additional habitat for wildlife.

Waste Reduction and Recycling

1. Composting:

- **Organic Waste:** Compost crop residues, manure, and other organic waste to produce nutrient-rich compost for soil amendment.
- **Vermiculture:** Use earthworms to break down organic waste into high-quality vermicompost, enhancing soil fertility.

2. Recycling and Reuse:

- **Farm Plastics:** Recycle or reuse farm plastics such as mulch films, greenhouse coverings, and irrigation tubing.
- **Packaging:** Use recyclable or biodegradable packaging materials for farm products to reduce environmental impact.

Social and Economic Sustainability

1. Fair Labor Practices:

- **Worker Rights:** Ensure fair wages, safe working conditions, and respect for the rights of farm workers.
- **Education and Training:** Provide ongoing education and training opportunities for farm workers to enhance their skills and knowledge.

2. Community Engagement:

- **Local Markets:** Sell your products at local markets, CSA programs, and farm stands to strengthen community connections and support the local economy.
- **Educational Outreach:** Host farm tours, workshops, and events to educate the community about sustainable farming practices and the importance of local agriculture.

Implementing Sustainable Practices

1. Planning and Goal Setting:

- **Assessment:** Conduct a sustainability assessment of your farm to identify areas for improvement and set achievable goals.
- **Action Plan:** Develop a detailed action plan outlining specific steps to implement sustainable practices, including timelines and responsible parties.

2. Continuous Improvement:

- **Monitoring:** Regularly monitor the progress of your sustainability initiatives and make adjustments as needed.
- **Learning:** Stay informed about new sustainable farming techniques, technologies, and research. Participate in workshops, webinars, and farmer networks to learn and share knowledge.

Conclusion

Embracing sustainable farming practices is essential for the future of agriculture. By improving soil health, conserving water, managing pests naturally, and promoting biodiversity, you can create a resilient and productive farming system. Sustainable farming not only benefits the environment but also enhances the economic viability of your farm and the well-being of your community. In the chapters ahead, we will continue to explore strategies and innovations to help you succeed in a dynamic agricultural landscape. Get ready to cultivate sustainability and make a positive impact on your farm and the world around you!

Chapter 20: Managing Farm Labor

Welcome to the chapter on managing farm labor! One of the most critical aspects of running a successful farm is managing your workforce effectively. Whether you have a small team or a larger crew, your farm's productivity and overall success depend significantly on the skills, motivation, and efficiency of your labor force. This chapter will guide you through the essentials of hiring, training, and retaining farm workers, as well as creating a positive and productive work environment.

Hiring Farm Labor

1. Identifying Labor Needs:

- **Assessing Workload:** Determine the number of workers you need based on your farm's size, type of crops or livestock, and seasonal demands.
- **Job Descriptions:** Create clear and detailed job descriptions outlining the responsibilities, skills required, and working conditions for each position.

2. Recruiting Workers:

- **Local Communities:** Reach out to local communities, agricultural schools, and vocational training centers to find potential workers.
- **Online Platforms:** Use online job boards, social media, and farm-specific recruitment websites to advertise job openings.
- **Seasonal and Migrant Workers:** Consider hiring seasonal or migrant workers for peak periods. Ensure you comply with all legal requirements and provide adequate support.

3. Interviewing and Selection:

- **Screening Candidates:** Review applications and conduct interviews to assess candidates' skills, experience, and compatibility with your farm's values and work culture.
- **References:** Check references to verify the candidate's work history and reliability.
- **Trial Periods:** Consider offering a trial period to evaluate the worker's performance and fit with your team before making a long-term commitment.

Training and Development

1. Orientation:

- **Introduction to the Farm:** Provide new workers with a comprehensive introduction to your farm, including its history, mission, and values.
- **Farm Tour:** Give a tour of the farm to familiarize new workers with the layout, facilities, and safety procedures.

2. Skills Training:

- **On-the-Job Training:** Provide hands-on training for specific tasks such as planting, harvesting, animal care, and equipment operation.
- **Workshops and Courses:** Offer access to workshops, courses, and certifications to enhance workers' skills and knowledge.

3. Continuous Improvement:

- **Feedback and Evaluation:** Conduct regular performance evaluations and provide constructive feedback to help workers improve and grow.
- **Mentorship:** Pair less experienced workers with seasoned employees to facilitate knowledge transfer and skill development.

Creating a Positive Work Environment

1. Communication:

- **Open Communication:** Foster open and transparent communication channels. Encourage workers to voice their concerns, suggestions, and feedback.
- **Regular Meetings:** Hold regular team meetings to discuss progress, address issues, and plan for upcoming tasks.

2. Safety and Well-Being:

- **Safe Working Conditions:** Ensure that your farm complies with all health and safety regulations. Provide proper training, equipment, and protective gear.
- **Wellness Programs:** Promote worker well-being through wellness programs, adequate breaks, and access to medical support if needed.

3. Team Building:

- **Social Activities:** Organize social activities and team-building events to strengthen relationships and boost morale.
- **Recognition and Rewards:** Recognize and reward hard work and achievements to motivate and retain your workforce.

Retaining Farm Workers

1. Fair Compensation:

- **Competitive Wages:** Offer competitive wages that reflect the market rates and the skills of your workers.
- **Benefits:** Provide additional benefits such as health insurance, housing, transportation, and paid leave to attract and retain workers.

2. Career Development:

- **Advancement Opportunities:** Create pathways for career advancement within your farm. Promote from within whenever possible.
- **Skill Enhancement:** Support workers in acquiring new skills and qualifications that can lead to more responsibilities and higher pay.

3. Job Satisfaction:

- **Meaningful Work:** Ensure that workers feel their work is valued and meaningful. Show appreciation for their contributions to the farm's success.
- **Work-Life Balance:** Promote a healthy work-life balance by managing work hours and providing flexibility when possible.

Legal and Ethical Considerations

1. Employment Laws:

- **Compliance:** Ensure compliance with all local, state, and federal employment laws, including labor rights, wage laws, and safety regulations.
- **Documentation:** Maintain accurate records of employment contracts, work hours, wages, and benefits.

2. Ethical Practices:

- **Fair Treatment:** Treat all workers fairly and with respect, regardless of their background, ethnicity, or role on the farm.
- **Worker Rights:** Uphold workers' rights and ensure a workplace free from discrimination, harassment, and exploitation.

Utilizing Technology and Tools

1. Labor Management Software:

- **Scheduling and Time Tracking:** Use labor management software to streamline scheduling, time tracking, and payroll processes.
- **Task Management:** Implement task management tools to assign and monitor tasks, ensuring efficient workflow and productivity.

2. Communication Tools:

- **Messaging Apps:** Use messaging apps or communication platforms to facilitate quick and effective communication among team members.
- **Training Resources:** Provide access to online training resources, tutorials, and videos to support continuous learning and development.

Conclusion

Managing farm labor effectively is crucial for the success and sustainability of your farm. By hiring the right people, providing thorough training, creating a positive work environment, and ensuring fair treatment and compensation, you can build a motivated and productive workforce. Remember, the key to successful labor management is communication, respect, and continuous improvement. In the chapters ahead, we will continue to explore strategies and innovations to help you thrive in a dynamic agricultural landscape. Get ready to cultivate strong relationships and a thriving team on your farm!

Chapter 21: Marketing Your Farm Products

Welcome to the chapter on marketing your farm products! Effective marketing is essential for connecting your farm products with consumers who value quality, sustainability, and local produce. Whether you're just starting out or looking to expand your market reach, this chapter will guide you through proven strategies, tools, and techniques to effectively promote and sell your farm products.

Understanding Your Market

1. Identifying Your Target Audience:

- **Demographics:** Understand the demographics of your potential customers such as age, income level, lifestyle, and dietary preferences.
- **Consumer Behavior:** Analyze consumer behavior to identify their buying habits, preferences for local produce, and interest in sustainable farming practices.

2. Market Research:

- **Surveys and Feedback:** Conduct surveys or gather feedback from existing customers to understand their needs, preferences, and satisfaction levels.
- **Competitor Analysis:** Research competitors to identify gaps in the market, unique selling points (USPs), and pricing strategies.

Developing Your Marketing Strategy

1. Branding Your Farm:

- **Farm Identity:** Define your farm's unique story, values, and mission. Create a compelling brand identity that resonates with your target audience.
- **Logo and Visual Identity:** Design a distinctive logo and visual elements that reflect the essence of your farm and its products.

2. Product Differentiation:

- **Unique Selling Proposition (USP):** Highlight what sets your farm products apart from others, such as organic certification, sustainable practices, or superior quality.
- **Value Proposition:** Clearly communicate the value and benefits of your products to potential customers.

Choosing Marketing Channels

1. Direct Sales:

- **Farmers' Markets:** Participate in local farmers' markets to directly connect with consumers, build relationships, and showcase your products.
- **Farm Stands and On-Farm Sales:** Set up farm stands or establish an on-farm retail space to sell products directly to customers.

2. Online Presence:

- **Website:** Develop a professional website that showcases your farm, products, and purchasing options. Ensure it is user-friendly and optimized for mobile devices.
- **E-commerce:** Implement an online store where customers can browse products, place orders, and arrange for delivery or pick-up.

3. Social Media Marketing:

- **Platforms:** Utilize social media platforms such as Facebook, Instagram, and Pinterest to share farm updates, product photos, recipes, and customer testimonials.
- **Engagement:** Foster engagement with your audience through contests, polls, live videos, and behind-the-scenes glimpses of farm life.

Promotional Strategies

1. **Content Marketing:**

   - **Blogging:** Write blog posts on topics related to farming, sustainable agriculture, recipes using your products, or farm life. Share these posts on your website and social media.
   - **Email Newsletters:** Develop an email list and send newsletters with updates, promotions, seasonal recipes, and upcoming farm events.

2. **Community Engagement:**

   - **Community Events:** Host farm tours, workshops, cooking classes, or tasting events to engage with the local community and educate them about your products.
   - **Partnerships:** Collaborate with local restaurants, schools, community centers, or other businesses to promote your products and reach a wider audience.

3. **Customer Relationships:**

   - **Customer Loyalty Programs:** Reward repeat customers with discounts, special offers, or exclusive access to new products.
   - **Feedback:** Encourage customer feedback and testimonials to build trust and credibility. Address any concerns promptly to maintain customer satisfaction.

Measuring Success and Adjusting Strategies

1. **Analytics and Metrics:**

   - **Website Analytics:** Use tools like Google Analytics to track website traffic, visitor behavior, and sales conversions.
   - **Social Media Insights:** Monitor engagement metrics, follower growth, and demographic data to evaluate the effectiveness of your social media efforts.

2. **Feedback and Adaptation:**

- **Customer Surveys:** Gather feedback from customers to assess satisfaction levels, identify areas for improvement, and gather ideas for new products or services.
- **Adaptation:** Stay flexible and adapt your marketing strategies based on market trends, customer feedback, and the performance of promotional campaigns.

Conclusion

Effective marketing is crucial for successfully selling your farm products and building a loyal customer base. By understanding your target audience, developing a strong brand, choosing the right marketing channels, and implementing strategic promotional tactics, you can effectively promote your farm products and differentiate yourself in the marketplace. Remember, marketing is an ongoing process of learning, experimentation, and adaptation. In the chapters ahead, we will explore additional strategies and innovations to help you grow and thrive in the competitive agricultural industry. Get ready to elevate your marketing game and showcase the best of what your farm has to offer!

Chapter 22: Managing Farm Risks

Welcome to the chapter on managing farm risks! Agriculture is inherently exposed to various risks, including weather extremes, market fluctuations, pests and diseases, and operational challenges. Understanding and mitigating these risks are crucial for ensuring the resilience and long-term success of your farm business. This chapter will explore key aspects of farm risk management, from identifying and assessing risks to implementing strategies that safeguard your farm's profitability and sustainability.

Identifying Farm Risks

1. Risk Categories:

- **Production Risks:** Risks related to weather conditions, pests, diseases, and crop failures that can impact yield and quality.
- **Market Risks:** Fluctuations in commodity prices, demand shifts, and changes in consumer preferences that affect marketability and profitability.
- **Financial Risks:** Challenges related to cash flow, debt management, interest rates, and access to financing.
- **Legal and Regulatory Risks:** Compliance with environmental regulations, labor laws, and zoning ordinances that could lead to fines or legal disputes.
- **Personal Risks:** Health issues, accidents, or personal liabilities affecting farm owners, employees, or family members.

2. Risk Assessment:

- **SWOT Analysis:** Conduct a SWOT (Strengths, Weaknesses, Opportunities, Threats) analysis to identify internal strengths and weaknesses, as well as external opportunities and threats.
- **Farm Audit:** Perform a comprehensive farm audit to assess vulnerabilities, prioritize risks, and develop a risk management plan.

Strategies for Risk Mitigation

1. Diversification:

- **Crop Diversification:** Plant a variety of crops to spread risk across different growing seasons, market demands, and environmental conditions.
- **Livestock Diversification:** Raise multiple species or breeds to diversify income streams and mitigate market and disease risks.
- **Income Diversification:** Explore additional revenue sources such as agritourism, value-added products, or farm workshops to reduce dependency on a single income stream.

2. Insurance and Risk Transfer:

- **Crop Insurance:** Purchase crop insurance policies to protect against losses due to adverse weather, natural disasters, or yield fluctuations.
- **Livestock Insurance:** Insure livestock against disease outbreaks, accidents, or theft to minimize financial losses.
- **Liability Insurance:** Obtain liability insurance coverage to protect against legal claims or lawsuits arising from accidents or property damage on your farm.

3. Financial Planning:

- **Budgeting and Cash Flow Management:** Develop realistic budgets, monitor cash flow regularly, and maintain sufficient liquidity to cover operational expenses and unforeseen costs.
- **Emergency Fund:** Establish an emergency fund to buffer against unexpected expenses or income disruptions, such as equipment breakdowns or market downturns.

Environmental and Resource Management

1. Conservation Practices:

- **Soil Conservation:** Implement soil conservation practices such as cover cropping, contour farming, and reduced tillage to prevent erosion and preserve soil health.
- **Water Management:** Adopt efficient irrigation systems, rainwater harvesting techniques, and water-saving practices to conserve water resources and mitigate drought risks.

2. Integrated Pest Management (IPM):

- **Biological Controls:** Use natural predators, beneficial insects, and crop rotation to manage pest populations and reduce reliance on chemical pesticides.
- **Monitoring and Early Detection:** Regularly monitor crops and livestock for signs of pests or diseases. Implement early detection methods to prevent outbreaks and minimize damage.

Operational Risk Management

1. Infrastructure and Equipment Maintenance:

- **Regular Inspections:** Conduct routine inspections and maintenance of farm buildings, machinery, and equipment to ensure safe and efficient operation.
- **Safety Protocols:** Establish safety protocols, provide training for employees, and promote a culture of safety to prevent accidents and injuries.

2. Supply Chain Management:

- **Supplier Relationships:** Build strong relationships with reliable suppliers and contractors to ensure timely delivery of inputs and services.
- **Diversified Sourcing:** Identify alternative suppliers for critical inputs to mitigate supply chain disruptions and price fluctuations.

Crisis and Contingency Planning

1. **Risk Response Plan:**

   - **Emergency Response:** Develop an emergency response plan outlining procedures for responding to natural disasters, fire outbreaks, or other emergencies.
   - **Communication Plan:** Establish communication protocols to notify employees, customers, and stakeholders during emergencies and ensure timely updates.

2. **Business Continuity:****

   - **Backup Plans:** Create backup plans and alternative strategies to maintain farm operations during crises, such as alternative marketing channels or temporary relocation of livestock.
   - **Review and Update:** Regularly review and update your risk management and contingency plans based on changing circumstances, new risks, or lessons learned from past experiences.

Conclusion

Effective farm risk management is essential for safeguarding your farm's sustainability, profitability, and resilience in the face of uncertainties. By identifying risks, implementing mitigation strategies, and preparing for emergencies, you can protect your farm business and ensure its long-term success. Remember, risk management is an ongoing process that requires proactive planning, continuous monitoring, and adaptation to evolving challenges. In the chapters ahead, we will continue to explore strategies and innovations to help you navigate the dynamic landscape of agriculture and thrive as a resilient farm entrepreneur. Get ready to cultivate resilience and protect your farm's future!

Chapter 23: The Future of Agriculture: Innovations and Trends

Welcome to the chapter on the future of agriculture! As we navigate the complexities of a changing world, agriculture continues to evolve with advancements in technology, sustainability practices, and consumer preferences. This chapter explores the exciting innovations and emerging trends shaping the future of farming, offering insights into how farmers can adapt and thrive in an increasingly dynamic agricultural landscape.

Embracing Technology in Agriculture

1. Precision Farming:

- **Remote Sensing:** Utilize drones, satellites, and IoT (Internet of Things) sensors to gather real-time data on soil moisture, crop health, and field conditions.
- **Precision Application:** Implement precision application technologies for targeted fertilization, pesticide application, and irrigation, optimizing resource use and minimizing environmental impact.

2. Robotics and Automation:

- **Autonomous Machinery:** Integrate robotics and AI (Artificial Intelligence) in farm operations for tasks such as planting, harvesting, and sorting produce, enhancing efficiency and reducing labor costs.
- **Weeding and Monitoring:** Use robotic weeders and automated monitoring systems to manage weeds and pests more effectively while reducing reliance on herbicides.

Sustainable Agriculture Practices

1. Regenerative Agriculture:

- **Soil Health:** Focus on building soil health through practices like cover cropping, no-till farming, and crop rotation to enhance soil fertility, water retention, and carbon sequestration.
- **Carbon Farming:** Implement practices that capture and store carbon in soils, such as agroforestry, rotational grazing, and perennial cropping systems.

2. Vertical Farming and Controlled Environment Agriculture (CEA):

- **Indoor Farming:** Utilize vertical farming techniques in urban environments and controlled environment agriculture (CEA) systems like hydroponics and aquaponics to produce crops year-round with minimal water and land use.
- **LED Lighting:** Optimize plant growth with LED lighting systems tailored to specific crop needs, providing energy-efficient and customizable light spectrums.

Consumer Trends and Market Opportunities

1. Demand for Local and Organic Products:

- **Farm-to-Table Movement:** Capitalize on the growing consumer preference for locally sourced, organic, and sustainably produced food products.
- **Certifications:** Obtain organic and other certifications to meet consumer expectations for transparency, quality, and environmental stewardship.

2. Alternative Protein Sources:

- **Plant-Based Proteins:** Respond to the rising demand for plant-based foods and alternative protein sources by diversifying crop production to include legumes, pulses, and specialty grains.
- **Cellular Agriculture:** Explore opportunities in cellular agriculture for producing cultured meat, dairy, and seafood products using biotechnology and fermentation processes.

Resilience and Adaptation

1. Climate Change Mitigation:

- **Adaptation Strategies:** Develop climate-smart agriculture practices to mitigate risks associated with extreme weather events, droughts, and changing growing seasons.
- **Water Management:** Invest in sustainable water management strategies such as drip irrigation, rainwater harvesting, and water-efficient crop varieties.

2. Blockchain Technology:

- **Traceability:** Implement blockchain technology to enhance supply chain transparency, traceability, and food safety, providing consumers with verifiable information about product origins and production practices.
- **Smart Contracts:** Use smart contracts for efficient and secure transactions between farmers, suppliers, and consumers, reducing administrative burdens and improving trust.

Collaboration and Knowledge Sharing

1. Industry Collaboration:

- **Partnerships:** Foster collaborations with research institutions, technology providers, and industry stakeholders to access cutting-edge innovations, research findings, and market insights.
- **Knowledge Networks:** Participate in farmer networks, online forums, and agricultural extension programs to exchange ideas, best practices, and lessons learned with peers and experts.

2. Continuous Learning and Adaptation:

- **Education and Training:** Stay informed about emerging trends, technologies, and regulatory changes through workshops, seminars, and online courses.
- **Adaptive Management:** Embrace a culture of adaptive management, where farmers continuously assess outcomes, adjust strategies, and innovate based on real-time data and feedback.

Conclusion

The future of agriculture is filled with promise and opportunity, driven by innovation, sustainability, and evolving consumer preferences. By embracing technology, adopting sustainable practices, responding to consumer trends, and fostering collaboration, farmers can navigate challenges and seize new opportunities for growth and success. As you embark on this journey into the future of farming, remember to stay curious, adaptable, and resilient. In the chapters ahead, we will continue to explore strategies and innovations to help you thrive in a rapidly changing agricultural landscape. Get ready to cultivate innovation and shape the future of agriculture!

Chapter 24: Navigating Farm Regulations and Compliance

Welcome to the chapter on navigating farm regulations and compliance! Running a farm involves more than just planting crops or raising livestock; it also requires adherence to a variety of local, state, and federal regulations. Understanding and complying with these regulations is crucial to avoid legal issues and ensure the smooth operation of your farm. This chapter will guide you through the essential aspects of farm regulations, offering practical advice on how to stay compliant in a friendly and approachable way.

Understanding the Regulatory Landscape

1. Types of Regulations:

- **Environmental Regulations:** These include rules related to water usage, pesticide application, waste management, and conservation practices.
- **Labor Laws:** Regulations governing farm labor include wage standards, working conditions, child labor laws, and worker safety.
- **Food Safety Standards:** Guidelines and standards for the production, handling, processing, and labeling of food products to ensure consumer safety.
- **Zoning and Land Use:** Local ordinances that dictate how land can be used, including restrictions on building structures and conducting certain activities.

2. Regulatory Agencies:

- **Federal Agencies:** The U.S. Department of Agriculture (USDA), Environmental Protection Agency (EPA), and Food and Drug Administration (FDA) are key federal bodies overseeing farm regulations.
- **State and Local Agencies:** State departments of agriculture, environmental protection, and health, as well as local zoning

boards and health departments, play significant roles in regulating farm activities.

Key Areas of Compliance

1. Environmental Compliance:

- **Water Usage:** Obtain necessary permits for water usage, implement water conservation practices, and adhere to regulations governing irrigation and water rights.
- **Pesticide Application:** Follow guidelines for the safe application, storage, and disposal of pesticides. Maintain accurate records of pesticide use.
- **Waste Management:** Develop waste management plans to handle animal waste, crop residues, and other farm by-products in an environmentally friendly manner.

2. Labor Compliance:

- **Fair Labor Standards Act (FLSA):** Ensure compliance with wage and hour laws, including minimum wage requirements, overtime pay, and recordkeeping.
- **Occupational Safety and Health Administration (OSHA):** Implement safety protocols to protect workers from hazards, provide necessary training, and maintain a safe working environment.
- **Migrant and Seasonal Agricultural Worker Protection Act (MSPA):** Adhere to regulations protecting the rights of migrant and seasonal workers, including housing, transportation, and disclosure requirements.

3. Food Safety Compliance:

- **Good Agricultural Practices (GAP):** Follow GAP guidelines to minimize food safety hazards during the production and handling of fresh produce.

- **Hazard Analysis Critical Control Point (HACCP):** Implement HACCP systems for identifying and managing food safety risks in processing operations.
- **Labeling and Packaging:** Ensure that all food products are accurately labeled with ingredients, nutritional information, and allergen warnings as required by the FDA.

Navigating Zoning and Land Use Regulations

1. Understanding Zoning Laws:

- **Agricultural Zoning:** Familiarize yourself with local zoning laws that designate areas for agricultural use and impose restrictions on non-farming activities.
- **Building Permits:** Obtain necessary permits for constructing new buildings, expanding existing structures, or making significant modifications to your farm property.
- **Easements and Rights-of-Way:** Be aware of any easements or rights-of-way that may affect your farm operations, such as access roads, utility lines, or watercourses.

2. Staying Compliant with Zoning Requirements:

- **Regular Inspections:** Schedule regular inspections and maintain open communication with local zoning authorities to ensure ongoing compliance with zoning regulations.
- **Recordkeeping:** Keep detailed records of all zoning-related activities, including permits, inspections, and correspondence with regulatory agencies.

Tips for Staying Compliant

1. Regular Updates:

- **Stay Informed:** Subscribe to newsletters, join farming associations, and participate in workshops to stay updated on regulatory changes and best practices.

- **Consult Professionals:** Seek advice from legal and regulatory experts to navigate complex regulations and ensure full compliance.

2. Documentation and Recordkeeping:

- **Maintain Records:** Keep accurate and up-to-date records of all compliance-related activities, including permits, inspections, training sessions, and pesticide applications.
- **Audit Preparation:** Regularly review your records and conduct internal audits to ensure all compliance requirements are met and to prepare for external inspections.

3. Training and Education:

- **Employee Training:** Provide ongoing training for employees on safety protocols, food safety standards, and regulatory requirements.
- **Continuous Learning:** Invest in continuous education for yourself and your management team to stay ahead of regulatory changes and industry best practices.

Conclusion

Navigating farm regulations and compliance can seem daunting, but it is a vital aspect of running a successful and sustainable farm business. By understanding the regulatory landscape, staying informed about changes, and implementing best practices, you can ensure your farm operates within the law and avoids potential legal issues. Remember, compliance is not just about avoiding penalties; it's about fostering a safe, sustainable, and responsible farming operation that benefits you, your workers, your customers, and the environment. In the chapters ahead, we will continue to explore strategies and innovations to help you succeed in the ever-evolving agricultural industry. Get ready to cultivate compliance and thrive in a regulated world!

Chapter 25: Building a Farm Brand and Story

Welcome to the chapter on building a farm brand and story! Creating a strong brand and compelling narrative for your farm is essential for connecting with customers, differentiating yourself from competitors, and fostering loyalty. In this chapter, we will delve into the key components of branding, how to develop your farm's story, and practical steps to effectively communicate your brand to the public.

Understanding the Importance of Branding

1. What is a Farm Brand?

- **Identity:** Your farm brand is the unique identity of your farm, encompassing its values, mission, and personality.
- **Perception:** It's how customers perceive your farm and its products, influencing their purchasing decisions and loyalty.

2. Benefits of a Strong Brand:

- **Recognition:** A strong brand makes your farm easily recognizable and memorable to customers.
- **Trust:** Consistent branding builds trust and credibility, making customers more likely to choose your products over competitors.
- **Loyalty:** A compelling brand story fosters emotional connections, encouraging repeat business and customer loyalty.

Crafting Your Farm's Brand

1. Defining Your Mission and Values:

- **Mission Statement:** Clearly articulate your farm's mission — what you aim to achieve and why your farm exists.
- **Core Values:** Identify the core values that guide your farming practices, such as sustainability, community involvement, or quality.

2. Identifying Your Unique Selling Proposition (USP):

- **Differentiation:** Determine what sets your farm apart from others. This could be organic certification, heirloom varieties, innovative farming techniques, or superior quality.
- **Value Proposition:** Communicate the unique benefits your customers receive by choosing your products.

3. Developing Your Farm's Personality:

- **Tone and Voice:** Decide on the tone and voice that best represents your farm – whether it's friendly and approachable, knowledgeable and authoritative, or fun and quirky.
- **Visual Identity:** Create a cohesive visual identity, including a logo, color scheme, and design elements that reflect your brand's personality.

Creating a Compelling Farm Story

1. Elements of a Good Story:

- **Authenticity:** Be genuine and honest in your storytelling. Share the real experiences, challenges, and triumphs of your farming journey.
- **Relatability:** Connect with your audience by sharing relatable stories and values that resonate with them.
- **Emotion:** Evoke emotions through your story, whether it's passion for sustainable farming, the joy of harvest, or the dedication to quality.

2. Storytelling Techniques:

- **Personal Narratives:** Share personal anecdotes about why you started farming, your family's history, or memorable moments on the farm.

- **Customer Stories:** Highlight stories from satisfied customers, showcasing how your products have positively impacted their lives.
- **Behind-the-Scenes:** Give a behind-the-scenes look at your daily farm operations, showing the hard work and care that goes into producing your products.

Communicating Your Brand and Story

1. Online Presence:

- **Website:** Develop a professional and user-friendly website that tells your story, showcases your products, and provides information about your farm.
- **Social Media:** Utilize social media platforms like Instagram, Facebook, and Twitter to share updates, photos, and stories. Engage with your audience through comments and messages.

2. Marketing Materials:

- **Labels and Packaging:** Ensure your labels and packaging reflect your brand's identity and tell your story. Include your logo, farm name, and a brief description of your mission and values.
- **Brochures and Flyers:** Create informative and visually appealing brochures and flyers to distribute at farmers' markets, events, and local businesses.

3. Customer Engagement:

- **Farm Tours and Events:** Host farm tours, workshops, and events to give customers a firsthand experience of your farm and deepen their connection to your brand.
- **Loyalty Programs:** Develop loyalty programs to reward repeat customers and encourage them to share their experiences with others.

Building and Maintaining Brand Loyalty

1. Consistent Communication:

- **Regular Updates:** Keep your audience informed with regular updates through newsletters, blog posts, and social media.
- **Transparency:** Be transparent about your farming practices, product sourcing, and any challenges you face.

2. Customer Feedback:

- **Encourage Feedback:** Actively seek feedback from your customers to understand their needs, preferences, and suggestions for improvement.
- **Respond and Adapt:** Respond to feedback promptly and use it to make informed decisions and improvements.

3. Quality Assurance:

- **Consistency:** Maintain high standards of quality in your products to ensure customers always receive the best.
- **Innovation:** Continuously innovate and introduce new products or services to keep your offerings fresh and exciting.

Conclusion

Building a farm brand and story is a journey that requires creativity, authenticity, and dedication. By defining your mission, developing a unique brand identity, and telling compelling stories, you can create a strong connection with your customers and stand out in the marketplace. Remember, your brand is more than just a logo or tagline – it's the heart and soul of your farm, reflected in every interaction with your customers. In the chapters ahead, we will continue to explore strategies and innovations to help you grow and succeed in the agricultural industry. Get ready to cultivate your brand and share your farm's story with the world!

Chapter 26: Sustainable Farming Practices

Welcome to the chapter on sustainable farming practices! As awareness of environmental issues grows, sustainable farming has become a priority for many farmers. Sustainable practices not only protect the environment but also improve farm productivity and resilience. This chapter will explore various sustainable farming techniques, their benefits, and how you can integrate them into your farm operations in a friendly and approachable manner.

Understanding Sustainable Farming

1. What is Sustainable Farming?

- **Definition:** Sustainable farming refers to agricultural practices that meet current food needs without compromising the ability of future generations to meet theirs. It emphasizes environmental health, economic profitability, and social equity.
- **Principles:** The core principles include conserving resources, reducing pollution, maintaining biodiversity, and supporting local communities.

Soil Health and Management

1. Importance of Soil Health:

- **Productivity:** Healthy soil is fundamental to farm productivity. It supports plant growth, improves water infiltration, and helps in nutrient cycling.
- **Biodiversity:** Soil health promotes a diverse microbial ecosystem, which enhances plant health and resilience.

2. Practices for Improving Soil Health:

- **Cover Cropping:** Plant cover crops like clover, rye, or vetch during off-seasons to prevent soil erosion, improve soil structure, and enhance nutrient content.

- **Crop Rotation:** Rotate different crops each season to break pest cycles, reduce disease risk, and improve soil fertility.
- **Reduced Tillage:** Minimize tillage to maintain soil structure, reduce erosion, and preserve soil organic matter.

Water Conservation Techniques

1. Efficient Irrigation:

- **Drip Irrigation:** Install drip irrigation systems to deliver water directly to plant roots, reducing water waste and increasing efficiency.
- **Scheduling:** Water plants during cooler parts of the day to minimize evaporation and ensure maximum absorption.

2. Rainwater Harvesting:

- **Collection Systems:** Set up rainwater collection systems to capture and store rainwater for irrigation during dry periods.
- **Storage:** Use tanks or ponds to store harvested rainwater and reduce reliance on groundwater or municipal water sources.

Integrated Pest Management (IPM)

1. Principles of IPM:

- **Prevention:** Focus on preventing pest problems through cultural practices such as crop rotation, sanitation, and habitat manipulation.
- **Monitoring:** Regularly monitor crops for signs of pests and diseases to detect issues early and respond promptly.

2. Control Methods:

- **Biological Control:** Use natural predators, parasitoids, or pathogens to control pest populations. For example, ladybugs can help manage aphids.

- **Mechanical Control:** Implement physical barriers, traps, or handpicking to reduce pest numbers without chemicals.
- **Chemical Control:** As a last resort, use targeted and least-toxic chemical options, ensuring minimal impact on beneficial organisms and the environment.

Biodiversity and Ecosystem Services

1. Encouraging Biodiversity:

- **Habitat Creation:** Plant hedgerows, wildflower strips, and native vegetation to provide habitats for beneficial insects, birds, and pollinators.
- **Agroforestry:** Integrate trees and shrubs into your farming system to enhance biodiversity, improve soil health, and provide additional income sources.

2. Ecosystem Services:

- **Pollination:** Support pollinators like bees by providing diverse flowering plants and reducing pesticide use.
- **Natural Pest Control:** Encourage beneficial insects that prey on pests, reducing the need for chemical interventions.

Waste Management and Recycling

1. Composting:

- **Organic Waste:** Compost farm waste such as crop residues, manure, and kitchen scraps to create nutrient-rich organic fertilizer.
- **Benefits:** Composting reduces waste, improves soil health, and lowers reliance on synthetic fertilizers.

2. Recycling and Upcycling:

- **Materials:** Reuse materials such as plastic containers, wooden pallets, and old equipment for new purposes on the farm.
- **Innovation:** Get creative with upcycling to reduce waste and save money, while also adding unique features to your farm.

Renewable Energy and Resource Efficiency

1. Renewable Energy Sources:

- **Solar Power:** Install solar panels to generate renewable energy for your farm operations, reducing reliance on fossil fuels.
- **Wind Energy:** Consider wind turbines if your farm is in a location with consistent wind patterns.

2. Energy Efficiency:

- **Equipment Maintenance:** Regularly maintain and upgrade equipment to ensure it operates efficiently and uses less energy.
- **Energy Audits:** Conduct energy audits to identify areas where you can reduce energy consumption and improve efficiency.

Community and Social Responsibility

1. Supporting Local Communities:

- **Local Markets:** Sell your products at local farmers' markets, co-ops, and community-supported agriculture (CSA) programs to support the local economy and reduce food miles.
- **Education:** Engage with your community by offering farm tours, workshops, and educational programs to promote sustainable practices.

2. Fair Labor Practices:

- **Worker Rights:** Ensure fair wages, safe working conditions, and respectful treatment for all farm workers.

- **Community Engagement:** Participate in community initiatives and collaborate with local organizations to address social and environmental challenges.

Conclusion

Embracing sustainable farming practices is not only beneficial for the environment but also enhances the resilience and profitability of your farm. By focusing on soil health, water conservation, integrated pest management, biodiversity, waste management, renewable energy, and community engagement, you can create a sustainable and thriving farming operation. Remember, sustainability is a journey, and small changes can lead to significant impacts over time. In the chapters ahead, we will continue to explore strategies and innovations to help you succeed in the ever-evolving agricultural industry. Get ready to cultivate sustainability and secure a brighter future for your farm and the planet!

Chapter 28: Embracing Technology in Modern Farming

Welcome to the chapter on embracing technology in modern farming! In today's agricultural landscape, technology plays a pivotal role in enhancing efficiency, sustainability, and productivity on the farm. From precision agriculture to advanced machinery and digital tools, this chapter explores how farmers can leverage technology to optimize operations and achieve better outcomes.

The Evolution of Technology in Agriculture

1. Precision Agriculture:

- **Remote Sensing:** Utilize drones and satellite imagery to monitor crop health, soil conditions, and field variability, enabling targeted interventions.
- **GIS and GPS:** Implement Geographic Information Systems (GIS) and Global Positioning Systems (GPS) for precise mapping, planning, and navigation within fields.

2. Smart Farming Equipment:

- **Automated Machinery:** Deploy autonomous tractors and harvesters equipped with sensors and AI for precise planting, spraying, and harvesting operations.
- **IoT (Internet of Things):** Connect sensors and devices across the farm to collect real-time data on weather, soil moisture, and equipment performance.

Benefits of Technology Adoption

1. Enhanced Efficiency:

- **Optimized Resource Use:** Reduce input costs such as water, fertilizer, and pesticides through data-driven decisions and precise application techniques.

- **Labor Savings:** Streamline workflows and minimize labor-intensive tasks with automated systems, freeing up time for strategic management.

2. **Improved Crop Management:**

- **Predictive Analytics:** Use data analytics and modeling to predict crop yields, disease outbreaks, and optimal harvest times, improving decision-making.
- **Prescriptive Insights:** Receive actionable recommendations for crop management practices based on real-time data and historical trends.

Sustainable Practices Enabled by Technology

1. Resource Conservation:

- **Water Management:** Implement drip irrigation systems and moisture sensors to deliver water precisely where and when it's needed, reducing waste.
- **Precision Application:** Apply fertilizers and pesticides only where necessary, minimizing environmental impact and runoff.

2. Soil Health and Conservation:

- **Cover Cropping and No-Till Farming:** Preserve soil structure and fertility by minimizing soil disturbance and enhancing organic matter with cover crops.
- **Variable Rate Technology (VRT):** Adjust seeding rates and inputs based on soil variability across fields, optimizing yields and reducing costs.

Overcoming Challenges and Adoption Barriers

1. Cost Considerations:

- **Initial Investment:** Evaluate the upfront costs of technology adoption balanced against long-term benefits and potential savings.
- **Return on Investment (ROI):** Calculate ROI based on increased productivity, reduced inputs, and operational efficiencies over time.

2. Data Management and Integration:

- **Interoperability:** Ensure compatibility and integration of different technologies and data platforms to streamline operations and data management.
- **Cybersecurity:** Implement robust cybersecurity measures to protect sensitive farm data and ensure privacy compliance.

Integrating Technology into Farm Management

1. Education and Training:

- **Continuous Learning:** Stay updated on technological advancements through training programs, workshops, and industry events.
- **Skill Development:** Equip farm operators and staff with the skills needed to effectively use and troubleshoot technology on the farm.

2. Adaptation and Innovation:

- **AgTech Startups:** Explore partnerships with AgTech startups and tech providers to pilot new technologies and innovations tailored to farm needs.
- **Feedback Loop:** Gather feedback from farm employees and stakeholders to fine-tune technology applications and maximize benefits.

Looking Towards the Future

1. **Emerging Technologies:**

    - **Artificial Intelligence (AI) and Machine Learning:** Harness AI to analyze big data sets and optimize farm operations, from predictive modeling to autonomous decision-making.
    - **Robotics and Automation:** Develop next-generation robotics for tasks such as selective harvesting, crop scouting, and soil sampling.

2. **Sustainable Innovation:****

    - **Circular Economy:** Explore opportunities in bio-based materials, renewable energy sources, and closed-loop farming systems to enhance sustainability.
    - **Climate Resilience:** Adapt farming practices and technologies to mitigate climate risks and build resilience against extreme weather events.

Conclusion

Embracing technology in modern farming is key to unlocking new efficiencies, sustainability practices, and productivity gains. By integrating precision agriculture tools, smart farming equipment, and data-driven insights, farmers can optimize resource use, improve crop management, and foster sustainable practices. As you embark on your technology journey, remember to evaluate options carefully, invest in education and training, and stay informed about emerging trends and innovations. In the chapters ahead, we will continue to explore strategies and technologies shaping the future of agriculture. Get ready to cultivate innovation and thrive in a technology-driven farming environment!

Chapter 29: Farm Diversification - Strengthening Your Agricultural Enterprise

Welcome to the chapter on farm diversification! Diversifying your farm involves expanding beyond traditional crops or livestock to include a variety of products, services, or activities. This strategy can enhance resilience, create new revenue streams, and strengthen your agricultural enterprise. In this chapter, we'll explore the benefits of farm diversification, different diversification options, and practical tips for successful implementation.

Understanding Farm Diversification

1. What is Farm Diversification?

- **Definition:** Farm diversification involves adding new enterprises, products, or services to your existing farming operation.
- **Purpose:** It aims to reduce dependence on a single commodity or market, spread risk, and capitalize on emerging opportunities.

2. Benefits of Farm Diversification:

- **Risk Management:** Minimize the impact of market fluctuations, weather extremes, and disease outbreaks affecting specific crops or livestock.
- **Income Stability:** Generate additional revenue streams throughout the year, reducing reliance on seasonal income fluctuations.
- **Resource Utilization:** Optimize land, labor, and infrastructure by diversifying activities that complement existing farming operations.

Types of Farm Diversification

1. Crop and Livestock Diversification:

- **Alternative Crops:** Grow specialty crops such as herbs, fruits, or niche vegetables that have high demand or unique market niches.
- **Livestock Products:** Expand into raising specialty livestock breeds or producing value-added products like cheese or artisan meats.

2. Value-Added Products:

- **Processing and Packaging:** Add value by processing raw farm products into packaged goods like jams, sauces, or dairy products.
- **Direct Marketing:** Sell directly to consumers through farm stands, farmers' markets, or online platforms, offering fresh produce, meats, or handmade crafts.

3. Agri-Tourism and On-Farm Experiences:

- **Farm Tours:** Offer guided tours, educational workshops, or agritourism activities that provide visitors with a glimpse into farm life.
- **Farm Stay:** Host overnight guests in farm accommodations such as bed-and-breakfasts, cabins, or camping sites, offering a rural retreat experience.

Planning and Implementing Farm Diversification

1. Assessing Market Demand:

- **Market Research:** Identify consumer preferences, trends, and potential niches for new products or services in your local or regional market.
- **Customer Feedback:** Engage with customers to understand their needs, preferences, and willingness to support diversified farm offerings.

2. Financial Considerations:

- **Budgeting and Financing:** Evaluate the costs associated with diversification initiatives, including infrastructure upgrades, equipment, and marketing expenses.
- **Financial Planning:** Develop a business plan outlining expected revenues, expenses, and return on investment (ROI) for each diversification venture.

3. Infrastructure and Resources:

- **Utilizing Existing Assets:** Assess existing farm infrastructure, equipment, and human resources that can be leveraged for new enterprises.
- **Investing in Upgrades:** Upgrade facilities, storage, and processing capabilities to meet quality standards and regulatory requirements for new products.

Managing Risks and Challenges

1. Production Risks:

- **Crop and Livestock Management:** Implement sound agronomic practices and animal husbandry techniques to minimize production risks and ensure quality.
- **Disease and Pest Management:** Develop integrated pest management (IPM) strategies and biosecurity measures to protect diversified crops and livestock.

2. Marketing and Sales:

- **Brand Development:** Establish a strong brand identity for new products or services, highlighting the farm's values, quality, and sustainability practices.
- **Market Access:** Explore distribution channels, partnerships, and direct marketing opportunities to reach target customers effectively.

Measuring Success and Adaptation

1. **Monitoring Performance:**

   - **Performance Metrics:** Track key performance indicators (KPIs) such as sales growth, customer satisfaction, and profitability for each diversification initiative.
   - **Feedback Loop:** Solicit feedback from customers, employees, and stakeholders to continuously improve products, services, and visitor experiences.

2. **Adaptation and Flexibility:**

   - **Responsive Management:** Remain agile and adaptable to changing market conditions, consumer preferences, and external factors affecting farm operations.
   - **Continuous Innovation:** Foster a culture of innovation and creativity to explore new ideas, technologies, and partnerships that support sustainable growth.

Conclusion

Farm diversification offers numerous opportunities to enhance resilience, profitability, and sustainability in today's dynamic agricultural industry. Whether through new crops, value-added products, agritourism experiences, or innovative services, diversifying your farm can open doors to new markets and strengthen relationships with customers. As you embark on your diversification journey, remember to conduct thorough research, plan strategically, and leverage existing resources effectively. In the chapters ahead, we will continue to explore strategies and innovations to help you thrive in the ever-evolving agricultural landscape. Get ready to cultivate diversity and build a stronger, more resilient farm enterprise!

Chapter 30: Financial Management for Farmers

Welcome to the chapter on financial management for farmers! Effective financial management is crucial for the success and sustainability of your farm business. This chapter explores key financial principles, strategies for managing farm finances, and practical tips to help you make informed decisions and achieve your financial goals.

Importance of Financial Management

1. Financial Stability:

- **Risk Management:** Proper financial management helps mitigate risks associated with market fluctuations, weather events, and unexpected expenses.
- **Long-Term Viability:** It ensures the farm's ability to sustain operations, invest in improvements, and adapt to changing economic conditions.

2. Planning and Decision-Making:

- **Budgeting:** Establishing budgets allows you to allocate resources effectively, plan for expenses, and set realistic financial goals.
- **Investment Decisions:** Make informed decisions about equipment purchases, infrastructure upgrades, and expansion based on financial analysis and projections.

Core Financial Principles

1. Cash Flow Management:

- **Monitoring Income and Expenses:** Track cash inflows and outflows regularly to ensure liquidity and avoid cash flow shortages.
- **Seasonal Variations:** Prepare for seasonal fluctuations in income and expenses by building reserves during peak seasons.

2. Debt Management:

- **Strategic Borrowing:** Use debt judiciously for productive investments such as land acquisition, equipment upgrades, or diversification projects.
- **Repayment Planning:** Develop a repayment schedule that aligns with your cash flow and profitability projections to avoid financial strain.

Budgeting and Financial Planning

1. Creating a Farm Budget:

- **Revenue Forecasting:** Estimate income from crop sales, livestock, and other farm products based on historical data and market trends.
- **Expense Categories:** Allocate funds for operating expenses, inputs (seeds, fertilizers, feed), labor costs, equipment maintenance, and overheads.

2. Financial Projections:

- **Profit and Loss Statements:** Prepare regular financial statements to assess profitability, identify cost-saving opportunities, and monitor financial performance.
- **Balance Sheets:** Track assets, liabilities, and equity to evaluate the farm's financial position and net worth over time.

Managing Risk and Insurance

1. Risk Assessment:

- **Risk Identification:** Identify potential risks such as crop failure, livestock disease, or market volatility that could impact farm income.

- **Insurance Coverage:** Consider crop insurance, livestock insurance, liability insurance, and other coverage options to protect against financial losses.

2. Emergency Funds and Savings:

- **Building Reserves:** Maintain emergency funds to cover unexpected expenses or revenue shortfalls during challenging times.
- **Diversification Benefits:** Diversify income streams and products to spread risk and enhance financial resilience.

Investment and Growth Strategies

1. Capital Investment Planning:

- **Asset Management:** Evaluate the lifespan and ROI of farm assets (equipment, buildings) to prioritize investments and replacements.
- **Expansion Opportunities:** Assess feasibility studies, market demand, and financial feasibility before expanding operations or diversifying into new enterprises.

2. Financial Efficiency:

- **Cost Control:** Implement cost-saving measures such as efficient resource use, bulk purchasing, and negotiation with suppliers to optimize profitability.
- **Value-Added Opportunities:** Explore value-added processing, direct marketing, or agritourism ventures to increase revenue and profit margins.

Tax Planning and Compliance

1. Tax Management:

- **Tax Deductions:** Take advantage of tax deductions for farm expenses, depreciation, conservation practices, and retirement contributions.
- **Accounting Practices:** Maintain accurate records, receipts, and financial documentation to ensure compliance with tax regulations and reporting requirements.

2. Professional Advice:

- **Consulting Experts:** Seek guidance from accountants, financial advisors, and agricultural specialists who understand the unique financial challenges and opportunities in farming.
- **Continued Learning:** Stay informed about changes in tax laws, financial regulations, and industry trends through workshops, seminars, and industry publications.

Conclusion

Effective financial management is essential for farmers to achieve profitability, sustainability, and long-term success. By implementing sound budgeting practices, managing cash flow effectively, and making informed investment decisions, you can strengthen your farm's financial health and resilience. Remember, financial management is a continuous process that requires monitoring, adjustment, and strategic planning to adapt to evolving market conditions and achieve your financial goals. In the chapters ahead, we will delve deeper into specific financial strategies and tools to support your journey towards financial prosperity in agriculture!

Conclusion: Cultivating Success in Farming

As we conclude this book on starting and growing a farm business, I want to extend my heartfelt wishes for your success on this fulfilling journey. Farming is more than just a profession; it's a way of life rooted in passion, resilience, and dedication to feeding our communities and stewarding the land.

Throughout these chapters, we've explored the essentials of farm planning, from choosing the right crops and livestock to understanding market dynamics, implementing sustainable practices, and mastering financial management. Each topic has been crafted to equip you with practical knowledge, insights, and strategies to navigate the challenges and seize the opportunities that come with running a farm.

As you embark on this venture, remember that success in farming often comes from a blend of tradition and innovation. Embrace new technologies that enhance efficiency and sustainability while honoring time-tested agricultural wisdom passed down through generations. Remain adaptable to change, responsive to market demands, and committed to continuous learning and improvement.

Beyond the practical aspects, farming is also about community. Engage with fellow farmers, local businesses, and consumers who value fresh, locally grown products. Build relationships, share knowledge, and contribute to the resilience and vibrancy of your agricultural community.

As you tend to your fields, orchards, or livestock, may you find joy in every season's harvest, pride in your contributions to food security, and fulfillment in nurturing the land for future generations. May your farm thrive as a beacon of sustainability, productivity, and excellence in agriculture.

With all my best wishes for your success and prosperity,

www.ingramcontent.com/pod-product-compliance
Lightning Source LLC
Chambersburg PA
CBHW071932210526
45479CB00002B/641